WE WANTED WINGS:
A HISTORY OF THE AVIATION CADET PROGRAM

Dr. Bruce A. Ashcroft
Staff Historian
HQ AETC/HO
2005

OFFICER CODE

<u>Duty</u> well performed, <u>Honor</u> in all things, <u>Country</u> before self.

AVIATION CADET HONOR CODE

Article 1: An Aviation Cadet will not knowingly make any false statement, written or verbal, while acting in any capacity, official or otherwise, or in any situation reflecting on the Aviation Cadet Corps or the Air Force.

Article 2: An Aviation Cadet will not take or receive the property of another person, or persons, under any conditions, without specific authority of that person or persons.

Article 3: An Aviation Cadet will not impart or receive any unauthorized assistance, either outside or inside the classroom or places of instruction, which would tend to give any Aviation Cadet unfair advantage.

Article 4: An Aviation Cadet will not quibble, use evasive statements, or technicalities in order to shield guilt or defeat the ends of justice.

Article 5: An Aviation Cadet will report any violation of honor by another Aviation Cadet of which he is witness or has unquestionable knowledge.

Article 6: An Aviation Cadet will not commit any act of intentional dishonesty which will reflect in any way on the honor and integrity of the Aviation Cadet Corps and the Air Force.

> **Officer Code and Cadet Honor Code both from brochure, "Aviation Cadet Knowledge," Preflight Training School, Lackland AFB TX, 1959.**

FOREWORD

Over the last 40 years, the history of the aviation cadet program has been all but lost in the shrouds of time. For most of the history of our nation's air service, the flying cadet program, as it was known until World War II, and the succeeding aviation cadet program, was the primary source for air force rated officers—pilots, navigators, and bombardiers.

Though America's contribution to World War I was minor compared to that of the European nations, many of the flying cadets who trained as pilots became the backbone of the Air Corps during the interwar period and of the Army Air Forces during World War II. During the massive training programs of the Second World War, perhaps 200,000 aviation cadets completed training as rated officers. Unfortunately, no conclusive evidence has been found to tell us how many cadets completed training. Also, aviation cadets trained as flight engineers, armament officers, meteorologists, photographic officers, and communications officers during World War II, though these ground duty programs did not survive beyond the war.

The start of the Cold War brought the wholesale integration of jet aircraft into the air fleet and the creation of an independent United States Air Force in 1947. The aviation cadet program, again, provided the lion's share of the pilots, navigators, and bombardiers needed during the Korean War. The onset of the new technological era, however, led the Air Force to abandon the aviation cadet program because most of those who entered the program had little or no college education. The service decided it wanted officers with college degrees, and the Air Force Academy, established in 1954, Officer Training School, begun in 1959, and the existing Reserve Officer Training Corps programs made the cadet program obsolete. The last aviation cadet to earn his pilot's wings completed training in 1961, while the last navigator pinned on his wings in 1965.

This study will spark the memories of the rapidly dwindling number of cadets who went through training during World War II and the Korean War, and we trust it will prompt them to share their experiences with our command History Office. For those readers who are unfamiliar with the history of the cadet program, this study will give them a better appreciation of the rich heritage forged by these men, the aviation pioneers who laid the groundwork for today's Unites States Air Force.

DONALD G. COOK
General, USAF
Commander

TABLE OF CONTENTS

List of Figures

Table:

Introduction

The aviation cadet program was the source of most rated officers until the late 1950s, yet few in the Air Force today know anything about the program. In the earliest years of military aviation, formal training was limited. The First World War changed that, and the Air Service instituted the flying cadet program that served as the prototype for the development of Air Force rated officers for over 40 years. Over time, the requirements for entry into the program changed as the needs of the service changed and as technology developed. In the buildup for World War II, a more expansive "aviation cadet" program reflected the changing nature of air force specialties. While the emphasis would still be on pilots, the program produced a large number of navigators and ground duty officers. During the First World War, only a handful of flying cadets trained for other-than-flying specialties.

Unfortunately, command histories and other documentary sources do not always clearly identify the number of cadets who went through the various training programs. Officer students, most often graduates of West Point, aviation cadets, enlisted pilot trainees, and foreign students were commingled in the training program and program statistics often did not provide enough detail to account for each of the different categories of trainee. Also, in general, the historical record focuses primarily on the pilot training program, the largest of the training programs in number of training bases employed and graduates, and provides less detailed coverage of aviation cadets in navigator training, and has minimal coverage of other cadet programs. Consequently, this study, too, covers the pilot training program in greater detail.

During World War I, the flying cadet applicant had to be under the age of 25, have at least 2 years of college, meet rigid physical standards, and be morally sound. By the end of the war, over 10,000 American pilots received training at 41 stateside bases or at airfields in Europe and Canada. During most of the interwar years, flying cadets could only train to be pilots, and the cadets represented the largest source of air officers in the Army's small air arm. During World War II, a conservative estimate is that 250,000 aviation cadets graduated from pilot, navigator, and specialized training programs. In the direst days of combat, high school graduates as young as 18 qualified for the program, so long as they met qualifying scores on aptitude and medical screening tests.

During the demobilization following World War II, the aviation cadet program shut down briefly as the surplus of pilots and navigators left or were forced from the service. Beginning in 1948, however, the newly established United States Air Force trained about 5,000 pilots annually, most of whom were aviation cadets. The creation of an Air Force Academy in 1955 led, eventually, to the notion that USAF officers must be college-trained. With its emphasis on non-degreed students, the aviation cadet program would end in 10 years. The Air Force discontinued aviation cadet pilot training in 1961, just two years after the Academy graduated its first class, and closed its navigator training program for aviation cadets in 1965.

This brief overview of the aviation cadet program is only the beginning of the story. On the larger stage, the many young men who graduated as rated officers and technical specialists played a significant role in the nation's defense during World Wars I and II, the Korean War, the Southeast Asia War, and much of the Cold War.

Aviation cadets at Randolph Field plan their day's flying missions, 1941.   (USAF Photo)

Chapter I - The Flying Training Program through World War I

America's first military pilots flew balloons during the Civil War, observing troop movements. By the spring of 1863, however, operations had ceased due to their limited value. Thirty-five years later, balloonists in Cuba helped direct troop movements and artillery fire during the assault on San Juan Hill that made the Rough Riders and Teddy Roosevelt famous. Following the Spanish-American War, balloon companies operated intermittently as the Army struggled to incorporate aviation forces into the force. In May 1902, the Army organized a balloon detachment at Fort Myer, Virginia, though there would be little activity over the next several years. In 1906, U.S. Army officers such as Majors Henry B. Hersey and Samuel Reber, Capt Charles deForest Chandler, and Lt Frank P. Lahm promoted the service. Hersey and Lahm won the Gordon Bennett balloon race that year, crossing the English Channel in the balloon "United States;" Reber and Chandler represented the U.S. Army in a balloon ascent in Massachusetts sponsored by the Aero Club of America. In 1907, aviation activities advanced along several lines. Of greatest significance, perhaps, the Army established an Aeronautical Division on 1 August and assigned three men—Capt Chandler, Cpl Edward Ward, and First-Class Private Joseph E. Barrett—to the new office. The USAF traces its history to the creation of this small organizational entity. During the year, the Army bought two new balloons, received funding to purchase its first dirigible, and asked for bids on the production of an aircraft, after two years of prodding by the Wright brothers.[1]

While historians today know that airplanes would quickly prove to be a more effective defense technology than dirigibles and balloons, in 1907 no one could predict with certainty what the future held. Specifications for the Army's first dirigible included the provision that the airship be capable of carrying two people with a combined weight of 350 pounds and at least 100 pounds of ballast. Army officials required the successful craft to sustain a speed of twenty miles per hour over a measured course that required the pilot to fly both with the wind and against it. Thomas Scott Baldwin, who had built a succession of dirigibles between 1904 and 1906, won the contract, which included a provision for the instruction of two men in operating the airship. The cigar-shaped "Signal Corps No. 1" was 96 feet long, with a walkway suspended below the envelope. By walking back and forth on the undercarriage, the crew could help direct the airship's ascents and descents. During its acceptance test in August 1908, the dirigible remained aloft for two hours. On August 15, an unnamed correspondent for the *New York Times* who observed the trials described the airship landing "like a demon from the sky, its motor spitting fire and its long gray gas bag outlined against the sky of dusk." Baldwin taught Lieutenants Frank Lahm, Benjamin Foulois, and Thomas Selfridge to pilot "Signal Corps No. 1." Foulois, the first to pilot the dirigible, "surprised every one by the aptness with which he handled the planes." The Army used its only airship to train pilots and as a test platform for aerial equipment until 1912 when it sold "Signal Corps No. 1." The Army did not purchase another dirigible until after World War I.[2]

The procurement of the Army's first airplane was more difficult. Specifications required the plane to carry two men, that it remain aloft for one hour, and that it travel at an average speed of 40 miles per hour. The entry in the Signal Corps' "Log of Wright's Aeroplane" for 20 August 1908, noted that the "Machine arrived." On the 25th, the engines arrived and mechanics mounted them on the plane. Orville made his first practice flight at Fort Myer, just outside Washington, DC, on September 3 and on the 16th shattered the world record for sustained flight by remaining in the air for thirty-nine

Wright Flyer at Fort Myer.
(USAF Photo)

minutes. The next month of flights, however, was marred by tragedy. Lt Thomas E. Selfridge, flying as a passenger, was killed and Orville injured in a crash on 17 October. Orville returned home to Dayton on 1 November, ending his first attempt to meet the government requirements. In June 1909, both brothers went to Fort Myer for a second try. On 18 June, the plane arrived at the field, and the brothers made four short flights on the 29th. By 27 July, they were ready for their official test. The plane circled the field 79 1/2 times and successfully met the Army requirements. Formal acceptance of the flying machine followed on 2 August. To fulfill their contract obligations, the Wrights trained two men, Lt Lahm and Lt Fredric E. Humphreys. Because the test field at Fort Myer was so small, the Army moved the training program to College Park, Maryland. Humphreys was the first to solo, after receiving 3 hours, 4 minutes, and 7 seconds of instruction.[3]

The War Department established its first requirements for qualifying as a "military aviator" in February 1912. The pilot needed to make a flight lasting at least 5 minutes in a wind of at least 15 mph and attain an altitude of at least 2,500 feet at some point during his flight. The certification required the pilot to carry a passenger so that the combined weight of pilot and passenger totaled at least 250 pounds, and with the passenger on board attain an altitude during flight of at least 500 feet, then successfully land within 150 feet of a designated point. The pilot candidate also needed to demonstrate mastery of the machine during an engine-off situation and, finally, conduct a military reconnaissance flight of at least 20 miles at an average height of 1,500 feet. Once a student passed the qualifying test, the Army considered him a graduate of its training program. Though most applicants for flight training were officers, the War Department also allowed enlisted men to fly. Sgt Vernon Burge, who earned his certificate in August 1912, was the first enlisted airplane pilot.[4]

In addition to using College Park, the Army established a succession of small training programs, typically of brief duration, at Fort Sam Houston, in San Antonio; Augusta, Georgia; and Texas City, near Houston. Schools in the Philippines and Hawaii produced a few aviators. The Army subsequently concentrated its training program at North Island, near San Diego. Early programs typically consisted of a short series of flights, measured in minutes, attuned to the specific aircraft at hand. Controls for the early Wright and Curtiss airplanes purchased by the Army varied dramatically, and there was no standardized training curriculum. The Wright-method of pilot training employed an instructor sitting beside the pilot trainee, what came to be called the "dual method." Glenn Curtiss, at his North Island school, let the students fly by themselves, as instructors observed. The students progressed from a series of "grasshopping" flights, barely skimming the ground, to increasingly lengthy and more difficult endeavors. In December 1913, the Army recognized the school in San Diego as the "Signal Corps Aviation School."[5]

During these first years of military aviation, Army pilots developed their own distinctive uniform insignia as a means of identifying themselves. War Department General Order Number 39, issued 27 May 1913, authorized air force pilots to wear a badge showing their qualification. The first design for what quickly became known simply as "wings" included a bar with the inscription "Military Aviator" on it with an eagle suspended below. Twenty-four men received the Military Aviator badge by the end of 1913. Legislation passed in June 1916 allowed the Army to differentiate among its airmen by creating Junior Military Aviators (JMA)

Military Aviator Wings, 1913.
(USAF Photo)

and Military Aviators. After qualifying as a JMA, the pilot had to serve three years as an aviation officer to become a Military Aviator. For each of the ratings, the airman faced a board composed of three experienced aviation officers and two medical officers that ruled on his fitness. In August 1917, the Army created two sets of wings for its pilots. The qualification badge designed for the Military Aviator looked very much like the pilot wings of today, except that it was made of felt with the artwork done in silver thread. The shield of the United States was centered between two wings. The Junior Military Aviator badge had only one wing at first but was re-made into a two-winged design later in the year. To identify Military Aviators, the Army added a star above the shield; the Army awarded a one-wing qualification badge to its aviation observers. The Army adopted what today, for the most part, is the Air

Force standard set of wings in January 1919. Over time the air force adopted separate wings to identify those who qualified in specialties, such as airship pilots, airplane observers, and balloon observers and, then, during World War II, as bombardiers, navigators, and aerial gunners, among many sets of wings.[6]

Passed into law in July 1914, House Resolution 5304 created the Aviation Section in the Army, recognizing the growing importance of military aviation. In addition to administrative, executive, and scientific staff, the law authorized the section to have 60 aviation officers and "aviation students" and 260 enlisted personnel. New students, selected from those unmarried lieutenants under the age of 30 who had been recommended for aviation duty by the Chief Signal Officer, had to serve at least one year to determine their fitness for duty. The legislation increased flying pay from 35 percent of base pay to 50 percent. Lieutenants who applied for duty in the Aviation Section faced a qualifying physical examination administered by two medical officers. The act also provided flight training for as many as 12 enlisted men, based on the discretion of the Army's aviation chief. Separate legislation provided $250,000 in funding for aviation-related operations and procurement, double the prior year's appropriation.[7] At virtually the same time, European nations committed millions of dollars to the development of their air forces as war erupted in August 1914. Germany appropriated $45 million; Russia, $22.5 million; and France, nearly $13 million for aviation by the end of the year. At the lower end of the spending spectrum, the British and Italian governments set aside about $1 million each for their air forces.[8]

The Punitive Expedition

The Mexican Revolution, begun in 1910, spilled over into the American Southwest periodically throughout the decade and provided the U.S. air force its first lessons in aerial warfare. The Benjamin Foulois-led 1st Aero Squadron made the first mass deployment of American aviation forces in March 1916. The squadron helped chase Francisco "Pancho" Villa through northern Mexico following Villa's 9 March raid on Columbus, New Mexico, during which 17 Americans were killed. The next day, the 1st, then based at El Paso, received orders to join a retaliatory force being organized by U.S. Army Brigadier General John J. Pershing to capture Villa. Flying underpowered Curtiss JN-3 aircraft, the squadron's pilots provided limited, but useful, services in support of Pershing's troops: observation, carrying mail and official correspondence, and aerial photography. By mid-April only two of the original eight airplanes remained in service, and the squadron returned to Columbus for refitting. Four new Curtiss N8 aircraft were in Columbus awaiting the airmen's return and, in May, the squadron received 12 Curtiss R2s. The new planes also proved generally unsatisfactory for deployment and use in Mexico, but they were the best available for immediate delivery.

JN-3s in the field during the Punitive Expedition, 1916. (USAF Photo)

Altogether, between 15 March and 15 August, the pilots of the 1st Aero Squadron completed 545 flights in the United States and in Mexico, covering nearly 20,000 miles. More importantly, perhaps, as additional planes arrived at Columbus, the squadron tested the operational capabilities of the craft and conducted bombing and gunnery exercises. Over the next several months, until the end of the campaign in February 1917, the 1st flew only a few operational missions into Mexico.[9]

By this time, the Army's training program at San Diego had become significantly more sophisticated. In 1914, the Army abandoned its pusher-model trainers because they were so dangerous. In an accident, the rear-mounted engines often broke loose and fell on the pilots. Glenn Martin became the primary supplier of trainer aircraft. By 1 July 1915, 30 officers, 12 civilians, and 185 enlisted men were assigned to the Signal Corps Aviation School. Curtiss JN-series aircraft, forerunner of the famous JN-4 "Jenny," which would become the most-used trainer of World War I, were introduced in the summer of 1915. The school also divided its training program into three different phases, identifying different types of aircraft for each phase.[10]

In lighter-than-air developments, the Army assigned Capt Chandler to Fort Omaha, Nebraska, to organize the "U.S. Army Balloon School" in November 1916. The four-month course taught the operation of tethered and free balloons, and by the end of January 1917, 6 officers and 93 enlisted men had been assigned as school staff. The physical requirements of students entering the balloon program were the same as those for students entering pilot training,

though there were no age or rank restrictions.[11]

World War I

Against this background, the nation's air forces, and military services in general, were unprepared for global war. When the United States declared war on Germany in April 1917, the air force consisted of 96 rated officers and 2 flying schools. A handful of the pilots had served with the Pershing Expedition, but most had no combat experience; no specialized schools for observers or bombardiers existed at that time. In May, Alexandre F. Ribot, the French Premier, cabled President Woodrow Wilson asking the United States to send 5,000 pilots and 53,000 mechanics to Europe for the war. The War Department adopted Ribot's request as the basis for its Air Service training program. By the armistice in November 1918, however, over 11,000 pilots, observers, and bombardiers had completed training and $1.7 billion had been appropriated for aviation.[12]

Cadet-Related Legislation

The Army appropriation enacted in May 1917 authorized the War Department to acquire the land needed to establish "aviation schools, aviation posts, and experimental aviation stations and proving grounds." The law also allowed for the recall to active duty of reserve officers and enlisted men. The next year's appropriation, approved on 9 July 1918, recognized aviation students as "cadets" and specifically prohibited the Army from barring from service men who were not "equipped with a college education" so long as they were qualified in all other respects.[13]

Cadet Training Programs

To establish a training program to meet the unprecedented need for pilots, American officials visited Canadian schools, principally the University of Toronto School of Military Aeronautics (SMA), and used them as models upon which to pattern Air Service schools. American SMAs opened at six universities—the University of California, Cornell University, the University of Illinois, the Massachusetts Institute of Technology, Ohio State University, and the University of Texas—beginning on 21 May 1917, and programs at Princeton University and the Georgia Institute of Technology began operations on 2 July. Using terminology from the Canadian schools, the Air Service called the students "flying cadets." Pilot candidates had to be under the age of 25 and must have completed at least 2 years of college or a scientific school program. In addition to age and education qualifications, applicants had to "be athletic, honest, and reliable." Nearly 40,000 applied for the flying cadet program, 22,500 of whom passed the

physical that qualified them for entry. The training regimen included initial training in ground schools, eight weeks long at the start of the war, then six to eight weeks of flying instruction at a primary flying school. At primary, students typically spent between 40 and 50 hours in the air before attempting to pass their pilot qualification test. Cadets received their commissions as second lieutenants and were rated as Reserve Military Aviators or Junior Military Aviators, depending on whether they had come into the service in a reserve status or if they had entered training while already on active duty. A one-month advanced flying program qualified the pilots for pursuit, bombardment, or observation duties. Altogether, airmen completed about 90 hours of flying time before being sent to the front.[14]

The curriculum at the Schools of Military Aeronautics, which came to be called ground schools, included instruction in subjects such as: marching and close order drill, the duties and responsibilities of military officers, signaling, map reading, aerodynamics and the theory of flight, the rigging and repair of aircraft, the theory of internal combustion motors and motor repair, the theory of combat tactics, the operation of machine guns, and the theory of observation. Later in the war, the SMA course of instruction was extended to 12 weeks and in October 1918 to 20 weeks. In these schools, learning basic military discipline and how to be an officer were as important as aeronautical studies. The Army used 25 flying training schools in the United States during World War I for basic flying instruction, with Kelly Field in San Antonio graduating the most students (1,666).[15]

In advanced training, the newly-minted officers could pursue one of three specialized training programs—pursuit (fighter), bombardment, or observation (also known as Army Corps). While the officer's interests were taken into account, the commanding officer of the

Flying cadets in the classroom at Kelly Field, 1918. The white band on the hat identified the students as cadets. (USAF Photo)

base at which the cadet had taken primary typically decided the specialty. Initially a somewhat ad hoc system, by August 1918 the Office of the Director of Military Aeronautics outlined the traits desired for each of the three tracks:

Pursuit being purely offensive, a pilot's first qualifications should be aggressiveness and youth. He should be physically quick and alert. Flying should come naturally and easily. He should never be of the heavy, slow-thinking type. He should have initiative and quickness of perception. For Army Corps work, a pilot should be mature, serious, persist[e]nt, pay attention to detail, and be interested in military tactics and man[e]uvers. For Bombing, the older pilots should be chosen. They should be determined, have a good sense of navigation and [be] expert at cross-country flying.[16]

Advanced training in bombardment took place primarily at Ellington Field, Texas; observation at Taliaferro Field, Texas; and pursuit training at a number of fields. To hasten getting pilots to the front, many American pilots received their advanced flying training in Europe. The best known of the European fields was Issoudun, France. By August 1918, there were more men in the training system than could be effectively processed, so the Army began to concentrate training into larger schools, to close those locations with less favorable weather, to simplify logistics, and to better standardize training.[17]

10

Among the flying specialties, pursuit was far and away the most popular due to the growing popular image of the "ace," a combat pilot who shot down five enemy aircraft. Heralded in newspapers and magazines, American pilots like Edward V. "Eddie" Rickenbacker and Frank Luke, the balloon buster, captured the imagination, and Baron von Richthofen, the "Red Baron," remains a popular figure even as this study is being written. Colonel Walter C. Kilner, Chief of the Air

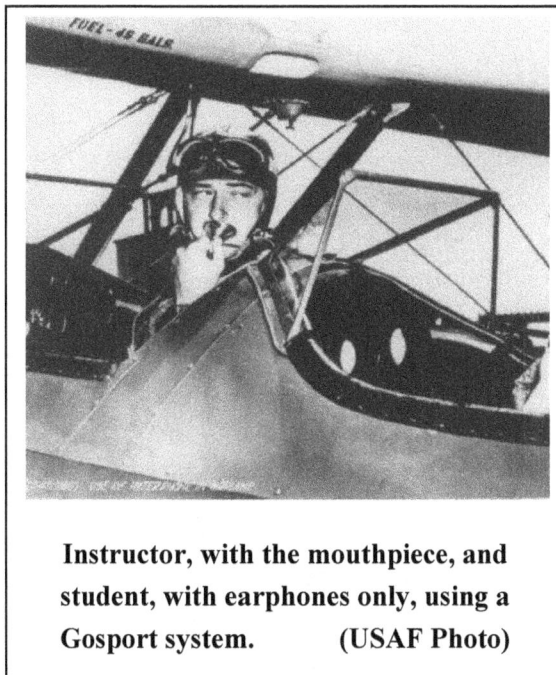

Instructor, with the mouthpiece, and student, with earphones only, using a Gosport system. **(USAF Photo)**

Service training section for the Allied Expeditionary Forces, lamented after the war that "the wide advertisement and publicity received by certain pursuit pilots probably did more harm to the other specialties of Aviation than any other one thing." Entering students dreamed of becoming an ace and looked down on the other specialties. Kilner

believed that observation was the most important of the three, even though it was the most unpopular. Observers, helping plot the location and movements of enemy forces, had the potential to affect the course of battle.[18]

Until the training program stabilized in the spring of 1918, Col Henry H. Arnold, serving as Assistant Director of Military Aeronautics in charge of the flying schools, described the training program more as a "state of affairs" than a "chain of events." While the Army planned to graduate 540 cadets per month during the first months of the training program, and increased the quota to 660 by the end of 1917, only 598 had completed ground school and primary training by 30 November 1917. A shortage of planes and training facilities, coupled with an initial reluctance to send cadets to Europe for training, handicapped the program.[19]

Most Army instructor pilots trained at Brooks Field, Texas, using the British method of training known as the Gosport system. Developed at the Gosport flying school, instructors shouted instructions into a funnel-shaped cone connected to a rubber tube that ended in earpieces inside the student's flying helmet. More important, perhaps, the Gosport method, which became the standard for training, stipulated that the same instructor work with the same students, usually five or six, throughout their course of instruction. This allowed

11

the students and instructor to build a rapport and avoided the need for students to adapt to varying instructional styles. The alternative was for students to progress through a succession of instructors in "stages," with specialized instructors providing different aspects of training.[20]

In July 1918, Dr. Edward L. Thorndike, who served on the Committee on Classification of Personnel in the Army, finished a report on physiological and psychological tests used to determine aptitude for flying. *U.S. Air Service* magazine, the "official publication of the Army and Navy Air Service Association," carried a modified version of this study as a three-part series between June 1919 and January 1920. Thorndike sought to quantify those factors that made the best pilots and found that 83 percent of those who entered training were successful in ground school. During flying training itself, the leading cause of failure was an "inaptitude for flying." The low attrition rate during the first phase of training came from having students who wanted to fly, were physically gifted, and were intelligent. The study found, "The men who had the best records over the lines were not, to any considerable extent, the men of more education, or higher salary, or greater leadership in the school or community." Success as a military aviator required specialized abilities, including a strong desire to fly and fight,

and a combination of courage, determination, and endurance. Successful airman needed a keen sense of balance, a talent for marksmanship, quick reflexes, alertness, and emotional stability or "nerve."[21]

V. A. C. Hermon, writing about the Air Service tests of aptitude for flying, commented that:

> *We were instructed to select men of good education and high character, men who were in every way qualified and fitted to become officers of the U.S. Army—a rather intangible set of specifications. We were constantly enjoined to remember that the flying officer was not to be an 'aerial chauffer,' but a 'twentieth century cavalry officer mounted on Pegasus.'*[22]

Both Thorndike and Hermon commented on the need to better predict those who would successfully complete both ground and flying training as a means of maximizing the efficiency of training. This refrain has echoed down through the years as the air force sought the best candidates for its flying training programs.[23]

Statistics compiled at the University of Texas (UT) showed that flying cadets who had completed at least three years of college performed best in the academic setting of the School of

Military Aeronautics. At the same time, the oldest cadets, those over 27 years of age, had the greatest difficulty. Looking at the success rate by previous occupation, SMA officials found that all 33 of the flying cadets who had previously been teachers passed the ground school course. Among the 69 cadets who reported being mechanics by profession, 16, or 24 percent, failed the course. This attrition rate was among the highest by occupation group, slightly higher than the 22 percent of clerical workers (60 of 240) who washed out, but significantly higher than the 4 percent of those who classified themselves as students (14 of 363) who failed the program.[24]

Cadet Life

Entering cadets at the UT SMA were organized into a "junior" wing and, during the first three weeks, they were not allowed to leave the school grounds. Instruction during this period consisted mostly of military drill, military law, and Army regulations; the students also underwent medical screening and received their inoculations for overseas duty. At the end of the three-week period, the cadets were tested on their knowledge. Students who passed the exam were allowed to continue in the SMA and entered the "senior" wing. Senior cadets were allowed passes to go out on Saturday night, and to mark their passage into the second phase of

training, the cadets wore a white band on their hats. The curriculum in the last five weeks was mostly technical, with only one hour per day devoted to military drill. At the end of the course, students faced another exam. According to the SMA history written in 1919, ground school was the place to eliminate those "who were morally, mentally, or physically unfit material to become flying officers."[25]

The cadets began their days at UT with reveille at 5:30 a.m. Classes met from 7-11 a.m. and again from 2-4 p.m. Instructors led calisthenics before lunch and drill from 4-5 p.m. After retreat at 5:50 p.m., the students ate and then were expected to study from 8-9 p.m. Taps sounded at 9:30. A writer for the UT *Alcalde* magazine commented that "the work is nearly all practical, very little theory being given." Classes in the use of machine guns, telegraphy and signaling, theory of flight, military law, astronomy, and map reading challenged the students.[26]

Clarence D. Chamberlin, one of the first airmen to fly the Atlantic Ocean, attended the SMA at the University of Illinois, then earned his wings at nearby Chanute Field. In his book, *Record Flights*, Chamberlin recalled the frenzied weeks of ground school and the "conglomeration of learning they asked us to digest to become flyers, and how much of it was useless effort." Life at Chanute entailed "flying as much as possible...drilling enough to keep us from forgetting entirely the stiff discipline of ground school, attending a few classes, and spending more and more time as rigging crews on airplanes in the hangars." After soloing, Chamberlin progressed to spirals and figure 8s, then stunt flying and acrobatics. Before taking the Reserve Military Aviator test, flying cadets spent a few days in formation flying and cross-country flights. Chamberlin received his commission and wings on 15 July 1918. The young pilot transferred to Camp Dick, near Dallas, and Wilbur Wright Field, Ohio, before staging at Hoboken, New Jersey, for shipment to the front. The war ended before Chamberlin could sail for Europe.[27]

Some flying cadets trained to be observers and bombardiers rather than pilots and a very few qualified for technical programs. At the University of Texas, for example, among the 4,663 SMA graduates, 26 cadets trained as observers, 16 as bombardiers, and 10 cadets became radio officers. Also, a small number of students at the SMAs had already received their commissions. Again, from the University of Texas, 77 officers qualified to be observers and 6 as adjutants. A couple of these specialized programs had shortened curricula—the officer adjutant course lasted only 5 weeks and the cadet observer course was 6 weeks long. The officers who attended the UT SMA wore their regular uniforms and lived in separate barracks from the cadets. Their classroom curriculum was devoted almost exclusively to technical subjects on the understanding that they did not need the basic military drill and instruction. A historical study prepared in 1944 mentioned an engineering program for cadets without giving any specifics. No conclusive data was found to indicate how many flying cadets trained for specialties other than pilot.[28]

Kelly Field cadets on the flight line, 1918. **(USAF Photo)**

Chapter II - The Flying Cadet Program During the Interwar Years

The nation dramatically reduced its military services following the armistice, and from 138,997 personnel assigned on 30 June 1918, Air Service manning fell to 9,358 as of 30 June 1920. As a percentage of Army strength, however, the Air Service fell only slightly over this period, from 5.8 percent to 4.7 percent. The number of personnel assigned hovered around the 10,000 mark through 1928, then began a steady rise before the precipitous pre-World War II build up beginning in 1939. Airmen represented just less than 8 percent of assigned Army personnel in 1928 and grew to 12 percent by 30 June 1939. After robust spending during the war, the 1920 fiscal year appropriation for aviation totaled only $28.1 million, and Army spending on its air arm averaged about $15 million until fiscal year 1928 when it again topped $20 million.[29]

Several signal events in air force history occurred during the interwar years. Bombing trials in the early 1920s vindicated the position of airpower advocates who believed that airplanes could sink top-of-the-line battleships and, accordingly, that the air force should play a key role in national defense. The 1926 Air Corps Act redesignated the aviation branch and provided recognition of the air arm as being co-equal with the infantry and artillery. The 1926 legislation also included a five-year expansion program that included the reconstruction of March Field, California; the creation of Randolph Field, Texas; and the concentration of flying training in the San Antonio area. The act increased the number of officers authorized from 1,247 to 1,650; the number of enlisted men grew from 8,760 to 15,000; and the act authorized the Air Corps to have up to 2,500 flying cadets in training. Central to this study, the 1926 legislation established the Air Corps Training Center (ACTC), first located at Duncan Field, adjacent to Kelly Field, and then at Randolph, to manage a consolidated flying training program. Through the 1920s and early 1930s, an average of 200 aviation cadets completed training each year.[30] Technology advances pushed airplane performance to ever greater speeds and endurance records.

Legislation Establishing the Grade of Flying Cadet

In January 1919, about 2,400 cadets remained in training, but the Air Service came to no firm resolution on what to do with them. Finally, on 30 May the director of the Air Service directed that all "regular and systematic training" for cadets be discontinued, except for those who agreed to work in the mechanics shops and care for aircraft. By the fall, Air Service leaders were ready to resume flying training and established a two-tier training program. By then, Congress had passed legislation creating the grade of "flying cadet," limiting the number on duty to 1,300. A year later, the National Defense Act raised the number of flying cadets to 2,500, but subsequent Army cutbacks reduced the number to only 500 in February 1921. Throughout the interwar years, only unmarried men who were citizens of the United States between the ages of 20 and 27 and high school graduates could qualify for the program. Further, applicants had to provide

**Army airship *C-2* at Brooks Field.
(USAF Photo)**

documentation of their good character; exacting physical examinations screened out many candidates.[31]

The Flying Cadet Training Program

Carlstrom and March Fields, California, hosted a four-month primary phase, and a three-month-long advanced training program would be conducted at Rockwell Field, California, for pursuit pilots. The Army tapped Dorr Field, Florida, and Ellington Field, Texas, for advanced bombing training and Post Field, Oklahoma, for its observation program. Aircraft that the service had used during World War I, the "Jenny" and "DH-4," were the mainstays for the primary and advanced stages, respectively. Just over 200 flying cadets reported to Carlstrom and March Fields in January 1920 for training. Difficulties precluded the Air Service from executing its advanced training programs as planned, however, except for observer training. Bomber and pursuit groups stationed at Kelly Field provided the advanced training because the designated fields were unable. In June 1922, the Air Service consolidated pilot training in San Antonio, using Brooks Field for primary training and Kelly for advanced. The Army had used Brooks Field for aircraft pilot training during World War I, then converted the base into a balloon and airship school in 1919. In 1922, the Army moved its lighter-than-air training program to Scott Field, Illinois, to accommodate the consolidation. The training program was extended to six months for each of the two stages. Instead of using different instructors for progressively advanced aircraft, the schools adopted the Gosport practice of allowing the same instructor to work with a small group of students throughout their program. As weather permitted, instructors conducted flying training in the mornings and classroom instruction in the afternoons. By 1923, the advanced school curriculum included night training in navigation and cross-country flights. After completing advanced training, cadets received their commission and wings. Though geographically close, the flying training programs were not consolidated under the same command until 1926.[32]

Between 30 June 1919 and 1 December 1926, 1,494 flying cadets entered primary flying school, with only 415 completing primary and advanced schools. For comparison, 934 student officers and enlisted men entered training during these years, with 378 graduating. Over the same period of time, 70 flying cadets trained at the Balloon and Airship School, with 30 graduating; 151 student officers and airmen entered LTA (referring to both balloons and airships) training, with 108 winning their wings.[33]

Kelly Field News Letters from 1920 and 1921 document the advanced training program of that era. One innovation heralded by the First Wing Operations Office in its 11 September 1920 newsletter included changing from the highly specialized training employed during World War I to a more general program. Officials believed that classroom training for both pursuit and bombardment officers should be almost identical and that the flying program should be very similar. "The peace time Air Service officer should be thoroughly familiar with the work of all branches of the service" and "able to lead a Pursuit flight on a special mission, conduct a bombing raid or an Artillery Shoot, make a mosaic or successfully cope with any of the other flying duties by day or night that the Air Service might be called upon to perform in time of war." Twenty-three cadets were in training, 7 specializing in pursuit and 16 in bombardment.[34]

The 11 September 1920 newsletter contained an outline of the weekly training schedule from the week beginning 30 August through the week of 11 October. During the first week, no flying was scheduled, but thereafter, students spent the morning hours, 8:30-11:30, flying, as weather permitted. Afternoon classes covered a wide range of topics, from lectures in radio, reconnaissance, and bombing to aviation theory and history. The week of 11 October saw the culmination of this phase of training, as students took tests in radio on Monday, gunnery on Tuesday, and artillery on Wednesday. The 26 March 1921 newsletter identified 18 flying cadets who had completed their pursuit training and were ready to go before the Cadet Examining Board "to be examined as to their mental, moral, physical, and professional qualifications to be appointed second lieutenants in the Reserve Corps." In addition to their flying training, the airmen received the training needed so they could lead a squadron.[35]

Charles Lindbergh, the first person to fly solo across the Atlantic and the most famous of the flying cadets who trained during the immediate post-war period, wrote extensively about his experience. He entered training at Brooks in March 1924. The 104 students filled the cadet barracks and some were billeted temporarily in the recreation hall. Describing the new underclassmen, Lindbergh wrote:

Graduation picture of Charles Lindbergh, Kelly Field, 1925. (USAF Photo)

We were a carefree lot, looking forward to a year of wonderful experiences before we were graduated as second lieutenants and given our wings. We had no doubt of our ability to fly although most of us had never flown before, and we had yet to get our first taste of the life of a flying cadet.[36]

By bed time that first night, apprehension had begun to set in as the upperclassmen filled them in on what to expect. During their first two weeks at Brooks, the cadets received inoculations, learned the basics of cadet etiquette, and began military training. Flying training began on 1 April. Lindbergh and five others were assigned to Sergeant Wilson for their flying training. Curtiss Jennies remained the primary trainer, though they had received more advanced engines since the end of World War I. Lindbergh felt fortunate to have the enlisted pilot—who held the record for flying time in the Army, some 3,300 hours—as his instructor. After about ten hours of dual instruction, the cadets were allowed to solo. Weekdays began at 5:45 a.m. and flying about 7. After lunch, classroom instruction lasted from 1-5 p.m. Study began after supper and lasted until bed check at 10 p.m.[37]

When the instructor decided a student should be washed out, he passed the student to the stage commander for a check flight. The stage commander could reinstate the student or pass him along to the chief check pilot for one last chance. If the student failed his last check flight, there was almost no chance of remaining in training. The last court of appeal was a board of officers known as the "Benzine Board."[38] Student officers who washed out of training were returned to the branch of the service from which they had transferred; cadet washouts returned to civilian life. Students could also be removed from flying training for misconduct or for academic deficiencies. Lindbergh wrote that there was no disgrace in washing out and that "with the washing out process our barracks became less congested." Of the 104 who entered primary, 33 graduated and went on to Kelly for advanced training.[39]

At Kelly, the pilot candidates flew De Havilland DH-4s, another workhorse from the First World War. The students were allowed a few days of flying on their own to get accustomed to the new airplane, then took a progress ride with an instructor. If their progress was satisfactory, the students moved to the next stage of training, if not, they faced the possibility of washing out. As Lindbergh worked his way through the stages, he felt as if he was constantly under observation. The rigor of observing reveille first thing in the morning and standing formation, however, were things of the past. In addition to flying training, the students spent two weeks at the gunnery range at Ellington Field; after returning to Kelly, they flew a few hours in the full variety of Air Service planes—pursuit, bombers, and observation. Later, the cadets were given their choice of specialties and finished their training program flying that type of aircraft. With so few pilots earning their wings, the Army, apparently, had little problem accommodating the students' requests. Lindbergh chose pursuit. At the end of the year-long training program, 18 of the original 104 class members received their wings and their commissions as second lieutenants in the Air Service Reserve Corps.[40]

The Air Corps Training Center

War Department General Order Number 18 issued on 16 August 1926 established, in name at least, an Air Corps Training Center, with Brig Gen Frank Lahm in command. Lahm located his headquarters at Duncan Field, which during World War II would be incorporated as part of Kelly Field. The ACTC charter included the coordination and management of training at the Air Corps Primary Flying School at Brooks Field, the Advanced Flying School at Kelly, and the School of Aviation Medicine at Brooks. Lahm appointed MSgt Herman Levy as the Sergeant Major for the training center and drew from the personnel at Kelly, Brooks, and

Duncan Fields to build his administrative staff.[41]

The Air Corps published a recruiting pamphlet in 1928 explaining the opportunities available in the air service. "Aviation presents a fascinating career to the young man of good education, sound health, and keen spirits," began the tract. The publication reviewed the military schools and the qualifications needed for entry into the programs. The airplane pilot's course began at the Primary Flying Schools, at Brooks and March Fields, on 1 July, 1 November, and 1 March each year. March Field, which had closed in 1922 with the consolidation of flying training in San Antonio, reopened in 1926 as part of the Air Corps' five-year expansion program. The students received about 75 hours of flying time and classroom instruction in airplane engines, navigation, machine guns, radio, and other academic subjects. The curriculum at the primary schools lasted eight months. The first class of students to enter training under the new curriculum started its program at Brooks on 1 July 1927; the first class at March began training on 1 November 1927. By the late 1920s, the Air Corps had replaced the long-lived "Jennies" with the Consolidated PT-1 "Trusty" in its primary and basic training program. During primary, students learned the fundamental operations of flight, including some acrobatics. In basic they graduated to formation flying, instrument flight during which the cadet's cockpit was hooded, night flying, and radio beam flying.

At the four-month long Advanced Flying School (AFS) conducted at Kelly Field, students could expect about 175 hours in the air. Here the flying cadets received specialized training in attack, pursuit, bomber, or observation aircraft. Curtiss A-3 attack aircraft, Boeing PW-9 and Curtiss P-1 pursuit planes, Keystone LB-5 and Martin NBS-1 bombers, and Douglas O-2 observation aircraft crowded the Kelly runways. Ground school at the AFS included subjects such as cooperation between air forces and ground troops, bombs and explosives, aerial photography, the principles of war planning, and the duties of junior officers.

Headquarters ACTC staff, Kelly Field.
(USAF Photo)

The Balloon and Airship School, completely independent of the flying schools, began training on 15 September each year and included about 175 hours in the air. Graduates of the lighter-than-air course were rated as both "airship pilot" and "balloon observer." Flying cadets for both LTA programs came from the enlisted corps of the Army, the National Guard Air Corps, students currently enrolled and graduates from Reserve Officer Training Corps (ROTC) units, and from civilian life. Applicants had to be unmarried male citizens of the United States between the ages of 20 and 27; they had to have completed at least two years of college or pass an exam covering equivalent material; and they had to be of excellent character and be able to provide documentation of that trait. Finally, they had to be of sound physique and in excellent health. At the time the pamphlet was published, the educational requirement had just been increased from having a high school diploma to two years of college. Air Corps officials expected this to help reduce the high attrition rate from training and maintain a high professional standard among the officer corps. The new requirement also simplified the evaluation of educational credentials. Base pay was $75 per month and flying cadets received an additional $1 per day for meals.

After the War Department accepted his application, the applicant reported to an examining board at one of about two dozen locations. Physicians at the examining board conducted a physical exam that eliminated up to 80 percent of the applicants. If the applicant could not provide a satisfactory college transcript showing that he had completed at least two years of classes, he had to take an education test. Subject matter covered a wide range of subjects, reflecting the college courses a student would have taken during his first two years of class—U.S. history, English grammar and composition, world history, geography, math (including advanced algebra, geometry, and trigonometry), and physics. From July 1928 to June 1939, about 1,500 applicants took the education test; 411 passed. Psychological screening, the third component of the initial entrance exams, was conducted at the two Primary Flying Schools. Based primarily on the research conducted during World War I, the psychological screening tested applicants for reaction time, orientation during simulated flying maneuvers, and a personality analysis.

At the completion of flying training, the cadets were discharged from the service, given their ratings, and commissioned as second lieutenants in the Air Corps Reserve. Depending on how much money the Air Corps had, some graduates were offered positions on active duty for a year. At the end of the year, the airmen took a competitive exam to fill existing vacancies in the Regular Army. Men who chose not to serve on active duty were assigned to reserve units with a two-week active duty obligation each year.[42]

Student and instructor with a PT-3, "Trusty." The PT-3 had a slightly larger engine than the PT-1, also called a "Trusty." (USAF Photo)

With the changes in the curriculum introduced by the ACTC staff, the graduation rate at Kelly Field's Advanced Flying School dropped significantly. Between July 1922 and February 1928, 865 students entered the training program, 16 were killed, 343 were eliminated, and 26 were held over

to the succeeding class. Only 55.5 percent of the students, 480, graduated. Beginning in 1928, after the revised schedule had been fully implemented, student attrition at Kelly dropped to less than 10 percent. Statistics from 1928 to 1935 showed that of 2,051 students who entered training, 36 were killed, 26 were eliminated, and 63 were held over. Almost 94 percent, 1,924 students, graduated. As a percentage of entrants, the number of students killed was virtually the same, 1.8 percent compared to 1.85 percent. In addition to the changed curriculum, school officials credited the change in entrance requirements for the flying cadet program, two years of college instead of just a high school diploma, for the sharp drop in eliminations. Even with the new education standard in place, a small number of applicants were able to enter the flying cadet program based on the education exam.[43] Kelly Field officials also cited improvements in classroom instruction as a contributing factor. Technical and theoretical subject matter had been reduced in favor of more practical subjects and motion pictures with sound had been added to the courses.[44]

As expected, attrition in primary and basic was significantly higher than that in advanced. Between 1931 and 1938, 54 percent of flying cadets successfully completed primary; of those who entered basic training, 88.7 percent graduated and went on to advanced training. Five students were killed in primary and 14 in basic.[45]

Randolph Field

In April 1927, General Lahm appointed a board of officers to find a site for a new consolidated training center. With its history, favorable climate, and mostly flat terrain, San Antonio was a leading candidate from the start. The field had to be large enough to accommodate 500 planes so that primary, basic, and advanced flying training could be conducted in one location. The San Antonio Chamber of Commerce offered several possible locations over the next several months, and the Army finally settled on "Site Q," located near the town of Schertz about twenty miles from San Antonio. San Antonio Mayor Charles M. Chambers brought Randolph Field to the San Antonio area by issuing city bonds totaling about $500,000 to buy the land for the new training field, and then using back city taxes to retire the bonds. Because the air city was outside the San Antonio city limits, there was great debate about the legality of using city money for the project. Chambers admitted the illegality of the move while stressing the need; San Antonio was competing against a host of communities across the United States for the flying school. The San Antonio city council passed a law making the transaction legal, and the mayor put together a committee of businessmen to pitch the city as the best location for the field. On New Year's Eve 1927, city leaders offered to donate the 2,300 acres of land near Schertz to the Army for the new training center. The Air Corps accepted the offer, and in February 1928 Congress passed the legislation needed to accept the offer.[46] Army officials named the newly authorized flying training base near San Antonio after Capt William M. Randolph, a native of Austin, Texas, who died in an airplane crash in February 1928. At the time, Randolph was serving on the selection committee to name the new airfield.[47]

Though other air force bases had longer histories, Randolph Field became the air force "city on a hill," the ideal "air city," from its inception. The base administration building, affectionately know as the "Taj Mahal" because it reminded someone of the famous building in India, served as a beacon, literally and figuratively, for military aviation during the years of the Great Depression. Even today, base officials call the field the "Showplace of the Air Force," and the Taj is considered the most photographed Air Force structure. World War II pilot John J. Hibbits recalled his train trip to San Antonio for induction into the pilot training program and somebody shouting, "There's Randolph!" A fellow student stopped in the middle of a story he was telling, another awoke from a nap with a start. Hibbits "was already across the aisle, peering through the window at the gleaming tower I'd seen so often in pictures. There it was, rising out of the Texas prairie, the West Point of the Air, the mecca of every lad in the three

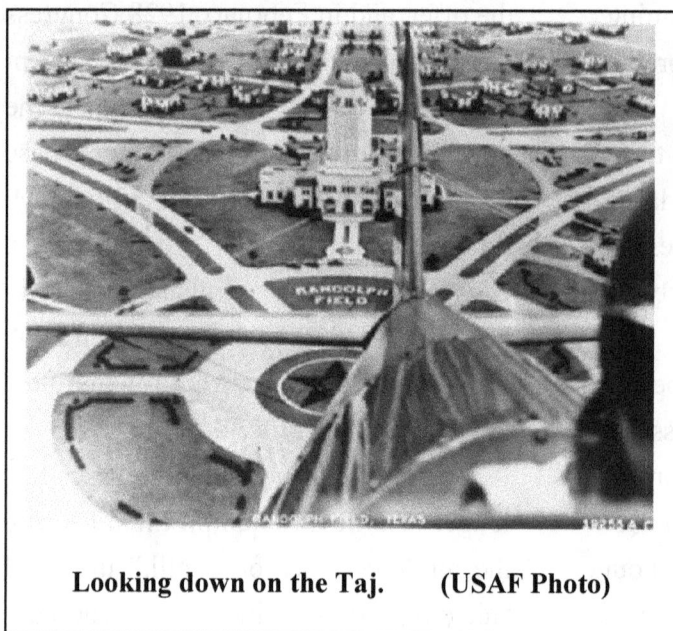

Looking down on the Taj.   **(USAF Photo)**

Pullmans."[48] The media embraced Randolph Field from the first, and air force leaders pushed their "West Point of the Air" forward at every possible occasion. When one recalls the history of the aviation cadet program, the years at Randolph Field represent the "golden age."

The base opened for business in October 1931 when the Air Corps Training Center headquarters moved to Randolph. The first flying training class consisted of 210 cadets and 99 student officers, relocated from schools at March Field and Brooks. Initially, officials thought Randolph would also house

Kelly Field's advanced training program, but concerns about having so many student pilots flying in a relatively confined area stopped the move. Just over 1,600 officers and enlisted men comprised the permanent party staff, with 111 aircraft assigned to the field. Select students from March and Brooks flew many of the planes to the new school.[49]

Like John Hibbits, airmen-to-be yearned to be accepted into the Randolph training program. The regimen was so difficult that school officials told the students that being "washed out" of flying cadet training should not be considered a failure. In the 24 classes that graduated between March 1932 and August 1939, slightly more than 50 percent of the 4,640 students who entered flying training were eliminated. Nineteen students either died in training-related accidents or were too seriously injured to continue. Men from all walks of life found their way into the program. The class of July 1937, for example, included 85 men who entered the flying cadet program right out of college and 49 who came to Randolph while on active duty. Six students had been school teachers, five had been clerks and another five general laborers, four each came from sales and civil engineers, and three had been coaches. Two lawyers, two chemists, two electricians, and two mechanical engineers joined the class, as did men who had jobs ranging from auto mechanic to musician. The students came from 40 states, the District of Columbia, Hawaii, Mexico, the Philippines, and Siam (now, Thailand). Texas, home to 25 cadets, was the most represented state, followed by Oklahoma (13), California (9), and Massachusetts (8). Most who aspired to join the air force did not even survive the initial physical examination; more failed the intensive physical and psychological scrutiny of the school's physicians. An article in the September 1932 issue of *Popular Science Monthly* noted that sometimes 90 percent of applicants failed the initial battery of physical exams. Of 2,124 applicants for the Army training program in 1934, 1,425, or 2 of every 3, were rejected during the initial screening process. Air Force physicians hoped to eliminate the training crashes that took such a toll on the World War I-era training programs by more carefully screening applicants. With the School of Aviation Medicine located at Randolph, the students provided a running laboratory experiment for the aviation medicine specialists at the field. Studies conducted during the 1930s showed that flying training program graduates weighed an average of 155 pounds and stood about 5' 9" tall.[50]

Everyday Life

Like Charles Lindbergh 10 years earlier, flying cadet program graduate Beirne Lay described the training program he experienced in his book, *I Wanted Wings* (1937). The first month at Randolph Field was known as "Hell Month," with its "Drill. Room inspections, rifle inspections, hazing. 'Yes, sirs. No, sirs. No excuse, sirs,' to the upperclassmen." Lay, who entered training at Randolph in 1932, described his initial reactions to a less-than-friendly reception at the West Point of the Air as a mix of bewilderment, stupefaction, and anger; "We all stood there at cramped attention and absorbed an oral punishment that was calculated to goad the most placid individual into a state of mental chaos." The constant barrage of bells added to the anxiety. "Clanging bells. 'Fall in for drill.' More bells. 'Fall in for fatigue,' two hours of picking up rocks and weeds from the parade ground and heaving them into G.I. cans. That damn bell again. 'Formal room inspection in three minutes.'" Lay came to realize that the purposes behind the endless drill and seemingly purposeless hazing was for the cadet to learn to "'take it,' to obey orders reasonable or otherwise without question, to withstand an overload on self-control, and to stay on our toes every minute." At the end of that first month, the cadets enjoyed open post and could travel to downtown San Antonio on weekends.[51]

For Lay and his fellow cadets at Randolph, the Advanced Flying School students were "gods—every one of them" and Kelly Field "'hallowed' ground." Lay went into bombardment training, expecting that the training on a multi-engine aircraft would later help him get a job with a commercial airline. Of the 204 who entered training at Randolph, 92 advanced to Kelly. Half chose pursuit training, 24 bombardment, 17 observation, and 5 attack. At Kelly, the buildings were old and the cadets had more freedom, but a bigger difference was that "realism was

Upperclassman putting "dodos" into a brace at Randolph Field. The underclassmen were called dodos, referring to birds that could not fly, until they had soloed. (USAF Photo)

creeping in." The basic reason Lay was in the flying cadet program was to learn how to use an airplane as a weapon. At the time Lay completed his training and received his commission, he was granted one year of active duty service. Four years later when his book was published, the flying cadet could expect five years of active duty service, with a $500 bonus at the end of that time. Graduates also had the option of reverting to an inactive status at the end of three years. Also, ACTC officials had cracked down on hazing and the horrors of hell month at Randolph Field, outlawing "unnecessary childishness and abuse of authority."[52]

In 1942, the Devin-Adair Company published a Federal Writer's Program history and guide to Randolph Field for the Work Projects Administration. Sections dealing with training rituals, cadet life, flying school "slanguage," maxims and proverbs, and versification speak to the culture of the school and the flying community. Seemingly bizarre activities at Randolph held deeper meaning as flying cadets learned the profession of aviation. At breakfast, underclassmen wore their flight goggles while eating to teach them how the goggles would limit their ability to see when they were flying. When making a turn while walking (or running), the underclassman extended his arms with the one on the side toward which he was turning pointed down, the other pointing up. This continually reminded him which way to bank an airplane to make a turn. In the barracks, the underclassmen were called upon to do squats (for physical training) and to hit their chests. The point at which they hit their chests was where their parachute rip cord ring would be found if they ever needed to bail out of an airplane.[53] Randolph's many rituals helped socialize the pilots-to-be and initiate them into an inner circle. Marching drills, which some considered out-of-date and irrelevant to those pursuing pilot training, contributed to the formation of the discipline needed by aviators. According to Lt Col I. H. Edwards, the Randolph field commander, "Precision and snap on the ground can and does promote similar characteristics in the air."[54] The flying cadets observed a strict protocol, clearly setting out the military hierarchy. Among the student pilots, West Point graduates and commissioned officers were recognized as "socially superior." The flying cadets, who would receive their commission if successful, did not mingle with the field's enlisted personnel.[55]

The creation of the ACTC, apparently, brought a stricter regimen to the Kelly Field program. An article in the 11 November 1931 cadet newsletter, *The Flying Cadet*, mentioned that the Advanced Flying School had "long been known as the Dodos heaven, but soon after the boys arrival the heaven turned to hell." Several changes had recently been instituted. The cadets marched in formation to meals and to and from the flight line and marched "double-time" to their physical training (PT) classes. The new administration also required the students to wear their dress uniforms to dinner. It came as a shock to some to learn "that they were still in the army, that they were still subject to restrictions and that there were still military schedules of training to observe." Academic classes met from 6:30-7:30 a.m., the students flew from 8-12, and attended classes again from 1-3. PT lasted from 3:30-4:30 p.m.[56]

Morse code training, probably the most-hated class cadets took. **(USAF Photo)**

The *Air Corps News Letter* regularly carried feature-length articles about Randolph Field and information about the flying cadet program. The 1 March 1937 issue, for example, described the evolution of the flying training program since the end of World War I, noting especially the increased complexity of the subjects taught because of the technological advances in aviation. The writer also commented about the much improved graduation rate: from less than 20 percent during the early 1920s to about 50 percent. Although the course length remained the same, one year, flying time had been increased by more than 60 percent. Two months later, the 15 May 1937 issue carried an article, "The Selection and Training of Flying Cadets," that spelled out in some detail the ACTC training program. The Air Corps Primary School was divided into three departments— Department of Flying, Department of Ground Instruction, and Military Department. In the Department of Flying, instruction was divided into two stages, each four months long. In the Primary Stage, students received approximately 70 hours of flying, learning basic operations and maneuvers in trainer aircraft specifically designed for the service. After eight to ten hours of flight with an instructor, students were ready to solo. The Consolidated PT-3 "Trusty," with a more powerful engine than the PT-1 "Trusty," was the first primary trainer used at Randolph. In 1936 the Stearman PT-13 "Kaydet" became the principal primary trainer. In the Basic Stage, students received about 118 hours of flight, including formation flying, instrument or "blind" flying, night flying, navigation, and radio beam flying. Students in Basic flew more

advanced aircraft than they had in Primary. The Randolph instructors initially used the Seversky BT-8 for basic training, with the North American BT-9 being introduced in 1935.

Students spent about 320 hours in the classroom during their eight months at Randolph Field. The Department of Ground Instruction, also known as the Academic Department, provided theoretical and practical instruction. Subject matter included the theory of flight, airplane construction, how to read the instruments to be found in the cockpit, and engines, including basic field repair. In "Buzzer Practice," instructors taught the sending and receiving of Morse code. Students also learned how to read maps and navigate. More mundane classes included military and civil laws and regulations related to aviation and military law and the manual of courts-martial. The Military Department provided the flying cadet training outside the formal classroom setting and gave him "the fundamental training and experience to enable him to qualify for duties of a junior officer of the Air Corps." The cadets were organized into a battalion, with upperclassmen and student officers performing duties such as Officer of the Day and Noncommissioned Officer in Charge of Quarters. Cadets received thorough indoctrination in infantry drill and ceremony, the responsibilities of guard duty, and military customs and courtesies.

At Kelly Field, students logged about 135 hours in the air. Academic instruction included lessons in many of the same topics introduced at Randolph, as well as cooperation with infantry and artillery, observation and reconnaissance, bombs and explosives, and aerial photography. Instructors taught the cadets about the organization of the Army and GHQ Air Force, the principles of war planning, and how to compose field orders, squadron orders, and flight orders. The cadets were given practical experience in the duties and

PT-13s on the ramp at Randolph Field.
(USAF Photo)

responsibilities of junior officers by being organized into a squadron, with duties as adjutant, engineer, supply, and mess officer. ACTC officials patterned their cadet program after those at West Point and Annapolis. The 15 May article claimed that in addition to turning out the best military airmen in the world, center officials were proud of the fine job graduates were doing in commercial aviation. And even if the graduates chose not to continue in an aviation profession, "we have helped build and develop finer American citizens." [57]

The Pre-World War II Expansion Program

The War Department created the General Headquarters Air Force in March 1935 to centralize command of combat air forces, at the same time maintaining the Air Corps primarily as an administrative and support headquarters. The first GHQ AF commander, Brig Gen Frank M. Andrews, reported directly to the Army chief of staff. Combat wings at Langley Field, Virginia, March Field, and Barksdale Field, Louisiana, provided a geographic distribution across the United States. Against increasing world conflict—the German annexation of territory in central Europe, Italy's conquest of Ethiopia, and Japanese aggression in the Far East—American leaders bolstered the nation's defenses. Personnel assigned to the air force topped 20,000 by 30 June 1938 and exceeded 50,000 two years later. By 30 June 1941, just months before the Japanese attack on Pearl Harbor that brought American involvement in World War II, over 150,000 men were assigned to the Army's air arm.[58]

The prewar expansion outgrew Randolph Field's capacity to train pilots. Major General Arnold, now the Chief of Air Corps, took the revolutionary step of asking private flying schools to conduct primary training and house, feed, and train flying cadets. The defense appropriation bill for January 1939 allowed Arnold to establish the first nine of what soon became fifty-six contract schools. Major players in the aviation industry, such as the Ryan School of Aeronautics in San Diego, responded to the call, as did aeronautic schools in major cities like Chicago and Dallas. In addition to the Ryan school, the first nine contract schools included: Alabama Institute of Aeronautics, Tuscaloosa, Alabama; Chicago School of Aeronautics, Glenview, Illinois; Dallas Aviation School, Dallas, Texas; Grand Central Flying School, Glendale, California; Lincoln Airplane and Flying School, Lincoln, Nebraska; Parks Air College, East St. Louis, Missouri; Santa Maria School of Flying, Santa Maria, California; and the Spartan School of Aeronautics, Tulsa, Oklahoma.

To produce pilots more quickly, the Air Corps reduced the training program from three, four-month phases to three, three-month phases. The Air Corps brought the contract school instructors to Randolph Field for training and indoctrination in the Randolph way of doing things. The Air Corps paid the contract schools a fee for each graduate, initially $1,170, and furnished the airplanes, a cadre of supervisory personnel, textbooks, and flying clothing for the students. The training program at the contract primary schools was nearly the same as that offered previously at Randolph. Col A. Warner Robins, the ACTC commander, reported to Air Corps headquarters in September 1939 that the civilian schools were performing better than expected. The accident rates at the schools were lower than the historical average at Randolph, and student elimination rates compared favorably.[59]

To help administer its training programs, in July 1940 the Air Corps divided ACTC into three commands—the West Coast Air Corps Training Center, with headquarters at Moffett Field, California; the Gulf Coast Air Corps Training Center, located at Randolph Field; and the Southeast Air Corps Training Center, at Maxwell Field, Alabama. Until 1943, Randolph Field maintained both instructor pilot and initial pilot training missions, but the two programs eventually overtaxed the Randolph facilities and the cadet program closed.[60]

Chapter III - From Flying to Aviation Cadets in World War II

The onset of World War II and the mobilization of millions of Americans reshaped the military services and, over time, was the crucible that proved the need for an independent air force in the United States.[61] As the international crisis deepened, the Army planned for ever-increasing numbers of pilots and fundamentally changed the flying cadet program. By 25 March 1941, almost nine months before the American entry into World War II, Congress authorized the production of 30,000 pilots annually, with over 100 training bases. The Army Aviation Cadet Act, Public Law (PL) 97 signed on 3 June 1941, created the grade of "aviation cadet" as a substitute for the grade of "flying cadet" established 22 years earlier. The law specified that "male citizens of the United States may enlist as aviation cadets, and enlisted men in the Regular Army may be appointed by the Secretary of War as aviation cadets." Cadets agreed to accept commissions as second lieutenants in the Air Corps Reserve and serve three years on active duty upon completion of training, unless released from service by the War Department earlier. At the end of the three-year period, the officers would be promoted to first lieutenant, again in the reserve. Base pay remained $75, as it had been in 1919, with a ration allowance of $1 per day. As under the previous legislation, cadets in flying programs qualified for flight pay; as a new perk, all aviation cadets received a $150 uniform allowance upon the successful completion of training. Reserve officers who were not offered commissions in the Regular Army and released from active duty were given a lump sum equal to $500 for each complete year of service. Another new benefit included government life insurance in the amount of $10,000 paid for by the government during initial training. After commissioning, the officers could continue the coverage at their own expense. Initially, the Army simply discharged aviation cadets who washed out of training. After 1 February 1942, the War Department required aviation cadets to enlist as privates in the Air Corps. Thereafter, if a cadet was eliminated from his training program, he reverted to the rank of private and could be reassigned according to the needs of the service. In November 1942, the active duty commitment was extended so that enlistments lasted for the duration of the war plus six months.[62] The three Air Corps Training Centers established in 1940 were realigned under the newly established Air Corps Flying Training Command (ACFTC) in January 1942.

Later, in July 1943, the ACFTC would become the Army Air Forces Training Command, responsible for both flying training and technical training programs.[63]

PL 97 did not set age or education qualifications, but the Army would adjust its requirements throughout the war according to its manpower needs. Less restrictive qualifications would allow more aviation cadets to enter training, more restrictive qualifications slowed the training pipeline. Almost immediately following the outbreak of World War II, the Army Air Forces (AAF), established in June 1941, dropped the requirement that aircrew trainees (pilots, navigators, and bombardiers) had to have college experience. Strict educational qualifications, however, remained throughout the war for those aviation cadets entering ground duty programs. In a 28 July 1941 memorandum, officials outlined three general classes, or tracks, for aviation cadets: "aviation cadet (pilot and bombardier)," "aviation cadet (navigator, non-pilot, flying)," and "aviation cadet (ground duty, as meteorologist and engineer)." Officials believed that each of the training programs would be of approximately the same duration. In actual practice, by the fall of 1941 the aviation cadet ground duty courses varied in length from 10 weeks to 9 months, while pilot training lasted about 7 ½ months. While there had been sentiment to withhold the commissions of the ground program graduates until they had completed the same length of service as pilot trainees, the Secretary of War ruled that aviation cadets should be commissioned upon the completion of their training programs, regardless of length.[64]

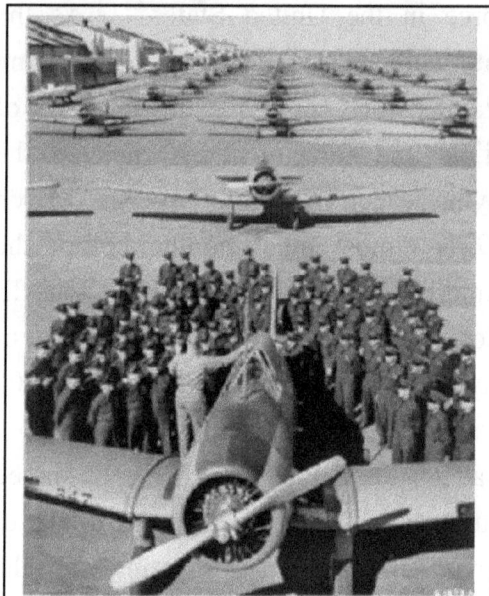

AT-6 "Texans" and aviation cadets. The AT-6 is the best known of the World War II-era training aircraft. (USAF Photo)

Recruiting brochures spelled out application procedures and outlined training programs. *Aviation Cadet Training for the Army Air Forces*, published in April 1943, for example, emphasized the urgency of the requirement—"Our nation's future depends upon command of the air. The future of freedom and liberty everywhere is in the hands of our youth." Men between the age of 18 and

26 could apply for aircrew positions, and those as young as 17 could apply with the consent of their parents. The April 1943 publication spelled out the need for meteorologists, noting that those up to age 30 could apply.[65]

By the end of World War II, nearly 200,000 men had been trained as Army pilots, along with 100,000 bombardiers and navigators, and another 20,000 ground duty officers. Unfortunately, the historical record does not accurately detail how many of these were aviation cadets. The AAF's Statistical Control Division did, however, provide a quarterly total of aviation cadets in training through World War II. In December 1941, only 17,614 cadets were in training, 16,733 of them in flying training programs and the rest in ground duty training. By the end of the next year, this had ballooned to 94,003 aviation cadets in training, 89,973 in flying programs and 4,030 in ground duty programs. The highest quarterly total, 114,336, was reached in December 1943, with over 109,000 of those aviation cadets in aircrew training programs. By December 1944, the training drawdown was well underway and 38,929 cadets were in training, only 133 of whom were in ground duty programs. To manage the cadet programs, the Office of the Chief of the Air Corps established a Flying Cadet Section early in 1940, renamed the Aviation Cadet Section with the passage of PL 97. Subsequently, the Headquarters of the Army Air Forces absorbed the Aviation Cadet Branch.[66]

Aircrew Qualification – The Classification Battery

To meet the ever-growing need for pilots and aircrew, in January 1942, the AAF abandoned the requirement that cadets in the flying programs had to have two years of college experience. Aviation Cadet Examining Boards administered a three-part battery of tests, in addition to a rigid physical examination, to help identify those who might make the best pilots, navigators, and bombardiers. The first part of what was being called the "classification battery" was the Aviation Cadet Qualifying Exam (ACQE). Designed to measure the candidate's comprehension, judgment, math skills, mechanical ability, alertness, and leadership qualities, the ACQE replaced the education test used in earlier years to measure the candidate's general knowledge and intellectual skills. Instead of being used to judge educational qualifications for those who did not possess the required school transcripts, the ACQE better predicted qualification for aircrew training and duty as an air force officer. A psychomotor test measured eye-hand coordination, reflexes, ability to perform under pressure, and visual acuity. Finally, the candidates underwent an interview with a trained psychologist.

School of Aviation Medicine staff testing a machine used to measure eye-hand coordination. **(USAF Photo)**

Stanines

After completing all three of the tests, officials scored the results on a nine-point scale. Called a standard nine, or "stanine," this composite score evaluated physical, psychomotor, and psychological attributes in an attempt to best match those entering the comprehensive training programs to the needs of the service. Initially, pilots and bombardiers required only the lowest stanine, 1, to qualify; in December 1942, air force officials raised this to 3. While pilot candidates only needed a low stanine score, reflecting the enormous need for aviation cadets to enter pilot training, navigator candidates typically required the highest stanine scores. Initially, navigator candidates needed a 5 stanine to qualify for training, and officials raised this to 7 in November 1943. Bombardier candidates needed a 6 stanine by mid-July 1943 to qualify for

training. In February 1944, AAF officials had pretty much leveled the stanines for the aircrew training programs. At that time, navigator candidates needed to score 6, bombardiers and pilots 5, to qualify for training. Students who had high stanines typically graduated at higher rates than those with lower scores. A faculty board at the training centers also considered the trainees' personal preferences in making the assignments. However, because 97 percent of the cadets wanted to be pilots, the classification procedure first took into consideration the type of aircrew training needed by the service, then the stanine score, and finally individual preferences.[67]

As spelled out in the *Aviation Cadet Manual* of 1942, those who would train as fighter pilots had to be between 64" and 69" tall, while other pilots could be as tall as 76". Navigators and bombardiers had to measure between 60" and 76" in height. The allowable weight was adjusted for height and weight, but no one who weighed more than 160 would be accepted into fighter pilot training and no one over 200 would be accepted into any of the aircrew programs. In July 1943, AAF officials relaxed the physical qualifications for aircrew members somewhat. Instead of requiring 20/20 vision, for example, 20/30 was acceptable so long as it was correctible to 20/20. In addition to eliminating the requirement for two years of college in January 1942 as noted before, officials instituted another significant change in requirements by allowing married men to apply for the aircrew programs.[68]

Between March 1942 and March 1944, aircrew classification centers processed 400,000 applicants, sending 260,000 to pilot training, 40,000 to navigator training, and 40,000 to bombardier training; the balance were eliminated from further officer training. By February 1944, the press to accept as many candidates as possible into the flying training programs was over. The War Department stopped accepting requests for transfer from officers and enlisted men in the Ground and Service Forces into flying training after 22 February. AAF Letter 35-66, dated 8 March, went further, curtailing the transfers of enlisted men into aircrew training programs, except crewmen who were returning from an overseas operational tour and AAF African-American enlisted men.[69] Men selected for ground duty programs did not go through the classification program.

Throughout the war, the AAF revised the ACQE: Test AC-10-A, for example, was officially put into use on 15 January 1942 and remained in effect for about 2 1/2 months before being replaced by Test AC-10-B. Version 10-B, in turn, was replaced by C, D, E, F, G, and H, with each iteration being used for about six weeks. The three-hour test included sections of vocabulary, reading comprehension, practical judgment, mathematics, alertness to recent world events, and mechanical comprehension. Scores in the mathematics section was a useful predictor of those who would be successful in navigator training, less so for pilots and bombardiers. Reading comprehension was important to the success of both pilot and navigator cadets; judgment scores were most predictive of pilot candidates' success. As the test matured, testing officials reduced the number of questions in the vocabulary and mathematics sections and increased the number of questions relating to current events and mechanical comprehension. Fewer than 40 percent of pilot candidates who scored 90 or higher on the test were eliminated from training, while almost 60 percent of those who scored lower than 90 washed out. For bombardiers, statistics were similar, but the test results varied significantly for navigator aviation cadets. Less than 20 percent of those who scored 90 or higher failed to complete training while over 50 percent who scored less than 90 washed out.[70] ACQE Test AC-12-I, introduced in July 1943, marked a significant change in the testing procedure. By the summer of 1943, the AAF recognized that a significantly increasing portion of aviation cadet applications were coming from men under 20 years of age who had no college experience. To compensate, officials developed a two-part test with less emphasis on educational material and more on perceptual skills. The first part of the test consisted of four, timed sections. The second part of the test was more nearly like the earlier versions of the ACQE.[71]

Aircrew Classification Centers

To facilitate the processing of the tremendous number of men needed for the aircrew training programs, Training Command established three aircrew classification centers in March 1942. Located in Nashville, at Kelly Field, and at Santa Ana, the classification centers were, essentially, collecting points where thousands of qualified candidates for aircrew training could be kept while awaiting their assignments. Here, would-be air force officers received their first uniforms and faced a series of tests, the classification battery. Officials expected to process as many as 205 applicants each day, six days a week, at Nashville, 154 at Kelly Field, and 102 at Santa Ana. This plan was based on the geographic distribution of the population of the United States. Eugene Fletcher, who arrived at Santa Ana on 7 January 1943, briefly described the physical, psychological, mental, and motor skills tests in his book, *Mister: The Training of an Aviation Cadet in World War II.* After receiving a series of vaccinations and having blood drawn, Fletcher

---

RANDOLPH'S AVIATION CADET CLASS 42-X

In 1942, as the shortage of flying instructors was becoming acute, the Gulf Coast Training Center experimented with a streamlined method of producing instructors. Class 42-X consisted of cadets selected from preflight school who showed an aptitude for teaching. They were sent directly to Randolph, bypassing the primary phase of pilot training. After graduating from this course, students were assigned to the Central Instructors School and various advanced schools to complete training.

The training period for 42-X was abbreviated, only 13 1/2 weeks, and some 4 1/2 weeks shorter than cadets usually spent in primary and basic. Despite the compressed training period, flying time exceeded that of the normal class by 48 hours. Of the 400 students who entered Class 42-X on 8 September 1942, 235 graduated. Although the elimination rate was high, it was still lower than what occurred normally in the primary and basic phases combined.

Based on the success of 42-X, Lt Col Harold A. Gunn, director of training at Randolph Field, recommended that students in preflight school be segregated into classes for single-engine and for multi-engine training and then given a compressed training program like that undertaken by 42-X. The single-engine course, starting on basic type training aircraft and continuing to advanced trainers and tactical fighter aircraft, would last 22 1/2 weeks. Multi-engine pilots would take a course of equal length, beginning with 9 weeks on primary type planes and then proceeding directly to twin-engine advanced schools. Flying Training Command officials disapproved the proposal.

"thought both arms would fall off; I had the feeling I had been kicked by a mule." After the embarrassment of the physical, he enjoyed the motor skills test as an

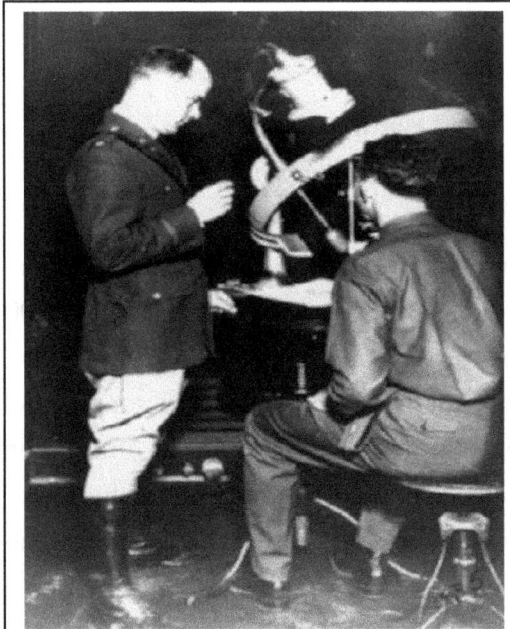

One of several vision tests, part of the medical screening process.

(USAF Photo)

exercise of man versus machine. The mental tests were a mystery, trying to decide what to tell "some sinister stranger what he sees in ink blobs. How far do you stretch reality before you fall off the deep end?"[72]

College Training Program

The College Training Program (CTP) was another expedient used during World War II to qualify potential aircrew members—pilots, navigators, and bombardiers—for training. With a backlog of nearly 100,000 applicants for the aviation cadet program, General Arnold developed the CTP to provide

additional training in math and physics, basic military indoctrination, and to keep the service volunteers employed while awaiting formal aviation cadet training programs. The program, which was in existence from March 1943 to June 1944, enrolled over 68,000 men at its peak in December 1943. Instead of processing through the Aircrew Classification Centers, those entering the aviation cadet program via CTP took qualifying education and medical exams at one of the 12 AAFTC basic training centers. Most of those who passed, were provisionally accepted as aircrew trainees and enrolled in classes at one of the 153 colleges affiliated with the program. The top 20 percent, however, bypassed college and were sent to the classification centers where they took the much more in-depth classification battery of tests to see which aircrew training programs they would enter. Those who failed the CTP qualifying tests were shipped to AAF technical training programs. Nearly 100,000 men entered the aviation cadet program through CPT. By the end of the program, the AAF's regularly constituted training facilities were producing enough aircrew members to meet the wartime need, combat casualties had begun to drop significantly, and the backlog of trainees had, for the most part, been eliminated.[73]

Flight Officers

At mid-war, some of the men completing aircrew training as aviation cadets did not receive their commission as second lieutenants. Instead, AAF Regulation 35-9 implemented the provisions of the flight officer act, allowing those who had lower grades in the training programs to receive a grade above the enlisted ranks but below that of commissioned officers. The men wore enameled rank insignia, a "blue pickle," so called because of the blue striping on the bars. Training Command and school officials established a cut-off grade for each graduating class that included academic marks, flying training, and military training records. Officials tried to reward men who might be good pilots, but who might not yet make good officers. A table prepared by the Aviation Cadet Branch in July 1944 showed that 91,545 men had been commissioned as second lieutenants and 7,458 had been designated as flight officers. At the same time, 16,891 had been commissioned as bombardiers, with 1,832 named flight officers; 17,026 navigators had been commissioned and 1,343 appointed as flight officers. Charles A. Watry, an aviation cadet in World War II who wrote a history of his experiences and the program in general, believed that cadets who only had a high school diploma and who were under the age of 20 were most likely to be appointed as flight officers instead of being commissioned as second lieutenants. The flight officer program was terminated in July 1947, two years after the war ended and just months before the Air Force became a separate service.[74]

Pilot Training

To support full-up production in anticipation of America's eventual entry into World War II, Training Command officials estimated that about 3,600 instructor pilots and 12,000 training aircraft would be needed to support the 30,000-pilot program. To meet the requirement, officials reduced the flying training program to three, nine-week phases in March 1942. Student pilots received about 200 hours in the air. Helping to streamline the program, initial military training, as well as administration of the classification battery, was moved into a preflight program. To accommodate this new phase of flight training, Training Command officials established three "Replacement Training Centers (RTC)," not to be confused with the classification centers, in January 1941 at already existing training facilities—Maxwell, Kelly, and Moffett Fields.[75] As with the classification centers, one of the RTCs supported each of the three Training Centers. Initially five weeks long, the preflight program was designed to provide trainees the fundamentals of military discipline and a baseline of academic information, and pilot candidates trained alongside bombardier and navigator candidates. In April 1942, the command adopted the term "Preflight School" instead of replacement training center, and by this time the Gulf Coast Training Center had established a separate preflight school for bombardier and navigator candidates; the other two centers had followed suit by June 1942. By spring 1944, AAF officials lengthened the preflight program to ten weeks.[76]

The pilot training program constantly evolved to match student throughput to the needs of the air force while maintaining quality. Briefly outlined, preflight training introduced students to the military and helped prepare them academically for the flying training program. The primary schools, all contractor-run operations, got the students into the cockpits, gave them the basic classroom knowledge they needed, and continued the process of making officers of the men. Basic schools introduced more involved flight maneuvers, and advanced schools prepared students for single-engine or multi-engine assignments. AAF officials cut the stages to 9 weeks in early 1942, then reverted to the 10-week schedule in mid-1944 as the critical need for pilots and aircrew members abated. At the height of the war effort, Training Command operated 6 preflight, 56 primary, 31 basic, 10 advanced single-

engine, and 22 advanced twin-engine schools. Each of the three flying training stages lasted 10 weeks during most of the pre-war buildup period. Most students eliminated from the program washed out in primary, where roughly one in four failed. Of the 233,198 men who made it through primary between July 1939 and August 1945, 202,986, or 87 percent of primary graduates, went on to successfully complete basic, and 193,440, 95 percent of basic graduates, received their wings.[77]

Books published by former aviation cadets, and magazines such as the Randolph Field *Form One*, detailed the day-to-day life of the aviation cadets; *Randolph Field: A History and Guide*, compiled under auspices of the Work Projects Administration at the start of World War II, captured the essence and color of the pilot training program. On Randolph Field, "the damndest noise you ever heard" woke the field's 500 aviation cadets for morning assembly at 5:45 a.m. After muster, the men rushed back to their barracks rooms to wash, shave, and make their beds before marching to breakfast. At each table in the messhall, an upperclassman served as the table commander, responsible for the conduct of those at his table. After breakfast, it was back to the dorm rooms to straighten up the rooms for inspection, then off to the classroom or the flight line. After classes, the cadets marched back to the barracks and subsequently spent an hour doing military drill or athletics. After supper, cadets remained in their room studying or engaged in extracurricular activities such as evening classes, chapel attendance, or sports. There was another room inspection before taps at 9:30 p.m. For those entitled to leave the base, "open post" on weekends began after dinner each Saturday; many journeyed to the cadet club at the Gunter Hotel in downtown San Antonio.[78] This basic scenario was repeated in cities and towns across the United States as the training program expanded.

At Randolph Field, itself, however, the aviation cadet program ended in March 1943 when the post became home to the Army Air Forces Central Instructors School (CIS). Commissioned offices selected for flying instructor duty attended the school, as did ground school instructors, officers selected for duty as commandant of cadets and administrative officers at Training Command bases, and civilian instructors at contract flying training bases. A local San Antonio newspaper lamented the change while recognizing the importance of the CIS:

When the last cadet marched out of Randolph the snappiest salute west of the Hudson left with him, along with eager faces turned to the skies, sabers flashing in the sun and columns of marching men wheeling on the parade grounds. The 'eager beaver' left too, and formal retreat, and short haircuts and the cadet 'slanguage' that he and he alone used.
Gone are the 'gigs' and demerits, the 'tours of duty,' rigid inspections and rigid cadet discipline; gone is the cadet, and Randolph will miss him.[79]

At the same time, the base assumed a more urgent task, training instructors. The CIS consolidated the training done at three regional schools into one program aimed at standardizing training throughout Training Command. Included in the training program was an attempt to develop a common standard among instructors for the evaluation of their students' military and professional bearing. CIS personnel also strove to instill high morale among those selected for instructor duty and to develop improved training methods and training aids. The *Training News*, published at Maxwell Field, carried an article in its 6 May 1944 issue, for example, extolling the pilot training manuals developed at the CIS. Instructors would use one set of books and students a complementary set. The article called the CIS "a clearing house for combat experience and new training ideas."[80]

40

As the aviation cadets, and other flying training students, went through training, they advanced through a series of progressively advanced training aircraft. While the planes used varied somewhat from school to school, the typical student started out on the PT-13 "Kaydet." Army officials introduced this biplane at Randolph Field in the late 1930s, and later versions of the same basic airplane but sporting different engines were designated PT-17, PT-18, and PT-27. The latter had an enclosed cockpit. Cruising speed for the primary trainer was just over 100 mph. The most widely used basic trainer was the Vultee

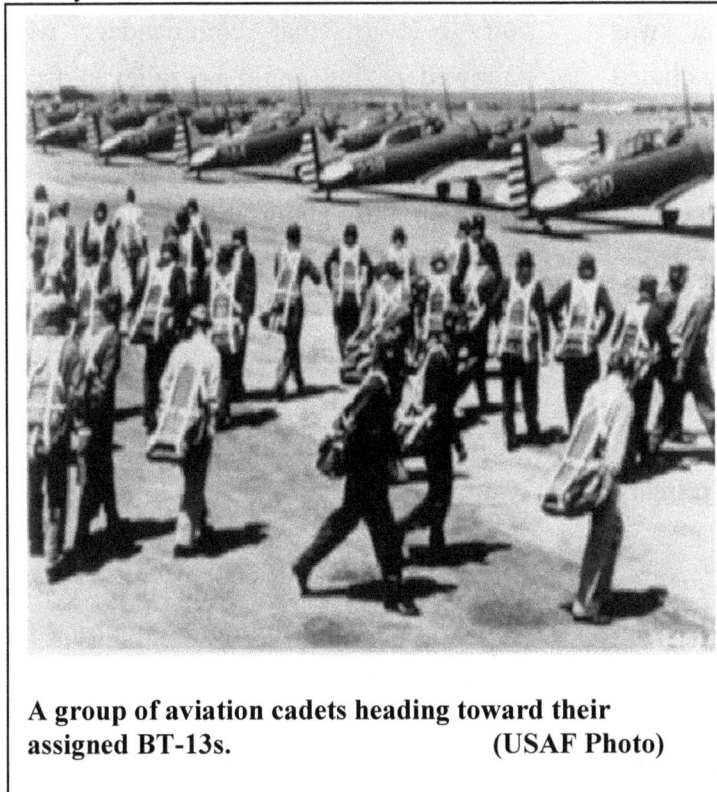

A group of aviation cadets heading toward their assigned BT-13s. **(USAF Photo)**

"Valiant," designated BT-13 and BT-15, depending on what engine was installed. Known to students as the "Vibrator" because of its flight characteristics, the

BT-13 was a monoplane that flew about 40 mph faster than the "Kaydet." This plane was considerably heavier than the primary trainer and employed landing flaps and a variable-pitch propeller for improved control. The "Valiant" also incorporated an electrical starter for the engine, navigation and landing lights for night flying, and a radio. The North American AT-6 "Texan," used among other aircraft in advanced training, was one of the most popular aircraft in aviation history. The U.S. Navy employed the AT-6, known as the "SNJ," in its flying training program, as did more than 30 Allied nations. British and Canadian students knew their version of the aircraft as the "Harvard." The monoplane aircraft had a maximum speed in excess of 200 mph and a retractable landing gear. Students flying the AT-6 repeated the mantra, "TMPFF" for trim, mixture, pitch, fuel, and flaps before take off and "UMPFF" for undercarriage, mixture, pitch, fuel, and flaps before landing.[81]

Aerospace historian, Bruce D. Callander, described his cadet experiences during World War II in a November 1990 article in *Air Force Magazine*. While the initial physical requirements remained high, he found the medical examiners lenient when he reported for testing weeks after the Pearl Harbor attack. Callander entered preflight training at the San Antonio Aviation Cadet Center (now Lackland AFB). The 10-week program was a "combination of basic training and Officer Candidate School, with a thin topping of West Point tradition." A few officers and NCOs ran the training program, but upperclassmen, theoretically getting experience in a command situation, administered most of the discipline. The differentiation among classes separated by scant weeks of entry made no sense and when 43-F became upperclassmen, hazing was outlawed. Preflight training included academic subjects such as physics, math, map reading, aircraft recognition, and code. Military drill and PT helped fill the days. Aviation cadets also stood guard duties, though Callander thought that there was nothing of interest for the potential saboteur or spy at the center. After preflight, Callander went to Victory Field, Texas, for his primary training.[82]

Publications such as *Compass Headings*, *Take-Off*, and aviation cadet orders or regulations sought to prepare cadets for the behavior that would be expected of them and potential disciplinary actions. The emphasis in such guides was duty and honor, reflecting the West Point code of conduct. While the pilot training programs paid homage to the "old days" and their Randolph Field heritage, the air force abandoned, in theory at least, one vestige of the aviation cadet program during World War II—hazing. Hazing, the harassment of underclassmen by those more senior, was the subject of Flying Training Command policy memoranda. FTC memorandum 50-0-2 of 15 May 1943, for example, prohibited upperclassmen from giving orders to underclassmen except as a result of official duties. "All forms of hazing, such as unauthorized calisthenics, stunts, etc., involving lower classmen will cease immediately." The memorandum went on to warn that commanders who allowed hazing could be relieved from their command.[83] In actual practice, hazing remained a time-honored tradition throughout the history of the aviation cadet program.[84]

There remained at the training centers an elaborate code of correct behavior, spelled out in cadet regiment regulations and handbooks. A Training Command letter sent to regional commanders in March 1942 noted that "the imposition of strict and unwaverable discipline" on entering cadets was necessary. However, "harsh, humiliating, or servile forms of hazing" were not in keeping with the objective of training future air force officers. FTC Memorandum 50-27-6, disseminated in July 1943, spelled out actionable training delinquencies and the demerits associated with each. Class I offenses, involving 11 or more demerits, were the most serious and included improper conduct that reflected on a cadet's character, intentionally destroying or damaging public property, and the use of narcotics. At the lowest range of the scale, Class VII offenses were assigned one demerit. These included lateness to a formation or class, articles in the cadet's room not being properly arranged, and dusty articles of equipment or clothing. After the accumulation of six demerits, aviation cadets walked one "tour," a one-hour march on Saturday afternoons for each demerit observed, or served one period of confinement for each excess demerit. Any single infraction involving seven or more demerits went to a delinquency board composed of commissioned officers for adjudication. The accumulation of sufficient demerits or single acts such as being absent without leave or involving significant character flaws warranted referral to a Faculty or an Academic Board for possible dismissal from the program.[85]

The December 1940 *Form One*, the aviation cadet magazine published at Randolph Field, contained an article from a cadet describing tours:

> *I am walking the ramp and my feet hurt. My 'tours' consist of carefully pacing up and down the length of an asphalt surface at the rate of one hundred and twenty steps per minute. I am very sure I don't like it. It is true the Lieutenant and I had a slight variance of opinion as to how thoroughly a rifle should be cleaned; it is true my originality in the matter of filling out Form 1 has been met with quite distinct disfavor. And again, I perhaps misjudged the importance of attending formations at the correct time.*[86]

The Randolph cadet calculated that he walked 3,600 inches each minute, 216,000 inches each tour. Altogether he walked five tours that Saturday afternoon, about 19 miles worth of 30-inch steps.[87]

Navigators, Bombardiers, and Observers

In addition to pilots, the World War II aviation cadet program provided training for navigators, bombardiers, observers, and ground duty specialties (see below). The 12,000-pilot program developed late in 1940, for example, included plans to produce 1,269 navigators, 778 bombardiers, and 900 observers. By the end of the war, almost 100,000 men had successfully completed flying training as navigators, bombardiers, and observers. As explained previously, qualifications and testing for these programs were similar

Navigators learning to "shoot the stars." (USAF Photo)

to those for pilot training, though bombardiers and navigators could be slightly shorter or taller—applicants at least 60" and not more than 76" tall were accepted into the program, provided they passed screening. After taking the preliminary physical and mental exams like other aviation cadet candidates, bombardier applicants, because they would be working with highly classified bombsights and other equipment, were questioned by members of the Aviation Cadet Examining Board to about their loyalty and patriotism. Stanine scores to qualify for bombardier and navigator training remained consistently higher than those needed to qualify for pilot training. An Air Corps paper from early 1942 specified that trainees would first complete a course in aircraft observation before specializing as either navigators or bombardiers. At the completion of

their programs, graduates would be rated as either Aircraft Observer (Bombardier) or Aircraft Observer (Navigator) and, like their pilot training counterparts, they would be commissioned as second lieutenants.[88]

As with the pilot production programs, the navigator and bombardier programs varied in length and content during the war. Pan American Airways provided a 15-week training program for air force navigators beginning in August 1940 at its Coral Gables, Florida, school. After Pearl Harbor, the Army Air Forces opened several navigation schools. Unlike the shortened length of pilot training, the curriculum was increased from 15 weeks to 18 in April 1943. Not until mid-1944 was the navigation course shortened to 16 weeks. By October 1944, Training Command had met its commitment for training navigators and slowed down the training pipeline. The course was extended to 20 weeks in December 1944 and then 24 weeks in June 1945. By 31 August 1945, the navigation schools produced 49,804 graduates. Early in the war the AT-7 "Navigator" was the most often used aircraft for navigation training, but the C-47 "Skytrain," better known as the "Gooney Bird," proved to be a better over-water trainer and was phased in at all of the schools.[89]

Bombardier training underwent the most radical change of the three major programs. What began as a 12-

week course at Lowry Field at the start of the war, grew into an 18-week program in June 1943 taught at eight schools. The impetus for the change was the decision to qualify bombardiers as navigators also, especially for the new B-29 bomber. As with navigator training, officials lengthened the bombardier course to 24 weeks as the war began winding down. Unlike other programs, there was no rigid elimination policy for bombardier cadets. So long as the student could hit the target, he might be continued in training until he met the academic requirements; each student was judged individually. Training officials had a more difficult time weighting the classification tests to best predict who would be a good bombardier, in part because it was not a pencil-and-paper job like that of the navigator, and the bombardier did not need the psychomotor skills of the pilot. Also, the bombardier had to work equally well as part of a team while being able to accept total responsibility for the success of a bombing mission. By 31 August 1945, Training Command schools produced 45,336 bombardier graduates. The AT-11 "Kansan" evolved into the standard bombardier trainer aircraft.[90]

One problem with recruiting men for navigator and bombardier training lay in the popularity of the pilot production programs and the publicity and glamour given to pilots. Navigators and bombardiers were seen as second-class citizens, a perception that continues today. As early as March 1942, Training Command officials addressed the issue of morale among aviation cadets training to be navigators and bombardiers. While first choice among training programs for most aviation cadet hopefuls was pilot training, the needs of the air force channeled many into other training programs. To minimize the disappointment, officials at the Santa Ana Replacement Training Center, for example, made "every effort…to educate and influence these cadets to be happy with their assignment." Training Command officials recommended to General Arnold that "an attempt should be made to publicize and glamourize bombardiers, navigators and gunners." Even prior to this suggestion, training officials at the Southeast Air Corps Replacement Training Center at Maxwell Field called bombardiers and navigators indispensable. A 4 October 1941 newspaper article about the non-pilot aviation cadet training programs at the field waxed eloquent about the nearly 300 aviation cadets training to be navigators and bombardiers then in school stating that "without them our flying fortresses would be eagles without talons or sense of direction." "The man who charts the course to the target or pulls the lever which 'lays the goose eggs' squarely, is as indispensable as the pilot." The book *Bombs Away*, written by Pulitzer Prize winning author John Steinbeck, and the Hollywood movie, *Bombardier*, much of which was filmed at the bombardier school at Albuquerque, helped publicize the bombardier program.[91]

Flight Engineers

Right at the end of World War II, AAFTC introduced a flight engineer program for aviation cadets. Like navigator, bombardier, and observer training, the program, specifically designed to produce flight engineers for B-29 bombers, was divided into a basic and an advanced phase. The 19-week basic course was offered at Amarillo Army Air Field (AAF), Texas, beginning in March 1945; the 10 weeks of advanced training was held at Hondo AAF. The course provided training in

AT-11 over a bombing range near Midland.
(USAF Photo)

B-29 airframe, systems, engine maintenance, the flight characteristics of the airplane, routine and emergency flight procedures, and aircraft operations. Students received 8 hours of flying time in a B-29. Through August 1945, 633 aviation cadets graduated as flight engineers.

Prior to the creation of the aviation cadet program, the Army trained engineering officers and enlisted men with B-29 maintenance experience to be flight engineers, but the increasing demand for B-29 aircrews opened the opportunity for cadets to enter this specialized field. The command histories did not indicate whether any screening tests were used to qualify aviation cadet applicants for the program, and the fuller story of the program was not told. Most likely, the classification battery used to initially qualify aircrew trainees was also used to provide intelligence, physical, and psychological screening. Apparently, the flight engineer program for aviation cadets was terminated by the end of 1945. No definitive training data was found and, presumably, students in training were offered the opportunity to simply leave the service after the war ended, as was the case with other training programs. When the flight engineer program resumed late in 1946, only aircraft maintenance officers were allowed into training.[92]

Ground Duty Programs

The Air Corps expansion in 1939 and 1940 led officials to expand the cadet program to include technical specialties. At a conference at Chanute Field in November 1940, university and military officials outlined curricula for maintenance engineering, armament, communications, meteorological, and photographic officers. Cadets, and officers who chose to retrain into these specialties, would enter maintenance engineering training at Chanute Field, armament and photographic training at Lowry Field, and communications training at Scott Field. Cadets learning meteorology studied at universities, most prominently the Massachusetts Institute of Technology, New York University, the University of Chicago, and the University of California at Los Angeles. Eventually, 27 colleges and universities provided meteorological training.[93]

An Air Corps Technical Schools brochure published in December 1941 described the four programs taught at Air Corps Technical Training Command schools. The engineering course lasted 120 academic days and provided training in airplane construction principles; maintenance, repair, and inspection of airplanes and associated tools and equipment; and the duties of an engineering officer. The armament course lasted 12 weeks and taught "the principles, operation, construction, adjustment, repair, inspection, and maintenance of all types of Aircraft Armament." The photographic course, which lasted 12 weeks "of 48 hours" each, according to the brochure, was designed to produce photographic laboratory commanders. As such, students learned about subjects as diverse as: "ground cameras" used to produce military identification cards, photographs for the historical record, for public relations, and for engineering studies; the development of aircraft camera films; how to interpret aerial photographs; and the optics and chemistry involved in the craft. The communications course lasted 16 weeks or 96 days. Radio fundamentals and operation constituted half of the curriculum. In November 1942, AAF officials established a technical school for meteorology at Grand Rapids, Michigan, to broaden the scope of training when it became apparent that the civilian schools could not produce the numbers of weather officers needed by the AAF. The new school increased

Aviation cadets studying meteorology.
(USAF Photo)

47

the number of meteorologists produced and provided training for the instructors who would teach meteorology at the aircrew training schools. The Grand Rapids facility was short lived, however, and by October 1943 was released by the AAF as requirements for meteorological officers had nearly been met. The training program transferred to Chanute Field, where the training of enlisted weather technicians had continued even after the Weather Training Center had been established in Grand Rapids. The college programs were phased out as the students completed their courses. In January 1943, the AAF began enrolling most of its maintenance, armament, photographic, and communications students in comparable programs at Yale University. Reading between the lines of the command history, one of the main advantages of the Yale programs was access to professional instructor staff.[94]

While ground-duty aviation cadets had to meet the same physical requirements as any regular Army Reserve officer, educational requirements for these technical specialists, in general, were strict. A study completed in November 1944 reviewed the changing requirements for entry ground duty training, beginning with a November 1940 baseline. At that time, the Air Corps limited those entering armament training to pilot training eliminees who had college degrees. Further, their commanding

officers had to recommend them for the programs. In January 1942, a month after the U.S. declared war, someone hoping to qualify as an aviation cadet in armament had to be a pilot training eliminee with two years of college, including a year of classes in physics, with his commander's recommendation for the program. Engineering officer applicants with a college degree in engineering received first consideration for the aviation cadet program in November 1940; those who were seniors in an engineering college got second priority. In January 1942, air force officials required three years of engineering at an accredited college or university to qualify. Initially, aviation cadets in photography had to be college graduates with a knowledge of chemistry or geology; in January 1942, three years of chemistry or geology study at a college or university sufficed. Applicants seeking appointment to the communications training program needed a college degree with "some knowledge of electricity or radio" in November 1940. By January 1942, the service accepted those who had completed two years of college with two years of engineering courses, but further required that applicants have an amateur radio license to qualify. No qualifications were listed in November 1940 for training as a meteorology aviation cadet, presumably because there were so few needed in the specialty at the time. By January 1942, however, officials recognized the contributions that weather experts could provide for mission operations and specified that those entering the program be college graduates who had concentrated studies in the sciences or engineering and who had completed math courses, including differential and integral calculus, and physics classes in heat and thermodynamics.[95]

In the ground duty programs, qualifications other than education were less stringent for aviation cadet applicants than for those seeking entry into flying duty programs. After January 1942, a married man could apply if he signed a statement saying that his family was self-sufficient without him. At that same time, the age qualification was relaxed to allow those as young as 18 to apply. By November 1942 the age limit was raised to 30, and in January 1943, the Air Corps decided to grant waivers for applicants up to age 38. Also in January 1943, the AAF allowed enlisted personnel to apply for the ground duty programs. Successful enlisted applicants were sent to a 9 weeks pre-technical training school at Boca Raton, Florida, to prepare them for the intense academic programs they were about to enter.[96]

By the end of 1942, the air force had built up a large backlog of qualified ground duty candidates awaiting training. By May 1943, AAF officials accepted no further applications from civilians, and in February 1944 stopped taking transfers from other branches of the Army. At the end of March, the aviation cadet examining boards stopped qualifying and accessioning new applicants for the ground duty programs. A snapshot as of 31 May 1944 showed that about 4,500 cadets remained in training. By that time, nearly 28,000 men had entered ground duty training, almost all of whom were cadets who came into the program from civilian life, from the enlisted ranks, or as washouts from one of the aircrew cadet programs. Only 717 of those who had entered ground duty training programs were officers retraining into one of the specialties. The May status report showed the following:

Specialty	Admissions	Graduates
Armament	3,419	2,505
Communications	7,815	5,715
Engineering	7,550	6,464
Meteorology	7,474	4,377
Photography	1,516	1,385

Most civilian students entered the program before the end of voluntary enlistment in December 1942, and most enlisted men who transferred into the programs came after this date. By September 1945, no ground duty aviation cadets remained in training.[97]

African-Americans and Women

One of the key social experiments of the World War II years was a fuller incorporation of African-Americans into the American military. The limited participation of the United States in the First World War had allowed the armed forces to exclude black Americans from entering pilot training, even though there were no laws specifically prohibiting the air force from accepting African-American men as flying cadets. Applications from black pilots were turned down because the Air Service was not forming separate squadrons for African-Americans and it was impossible to have integrated units. Throughout the interwar years, the air force continued to exclude African-Americans until Congress passed Public Law 18, General Arnold's Air Corps expansion act, in April 1939. The law included a provision for one civilian pilot training school for blacks, with the expectation that African-American pilots who passed the course would receive reserve commissions. Two months later, President Roosevelt signed a law establishing the Civilian Pilot Training (CPT) Program that allowed six historically black colleges to offer programs to African-Americans. In addition to the well-known Tuskegee Institute near Montgomery, Alabama, the West Virginia State College for

Negroes, Howard University, Hampton University, the Delaware State College for Colored Students, and the North Carolina Agricultural and Technical College hosted CPT programs.

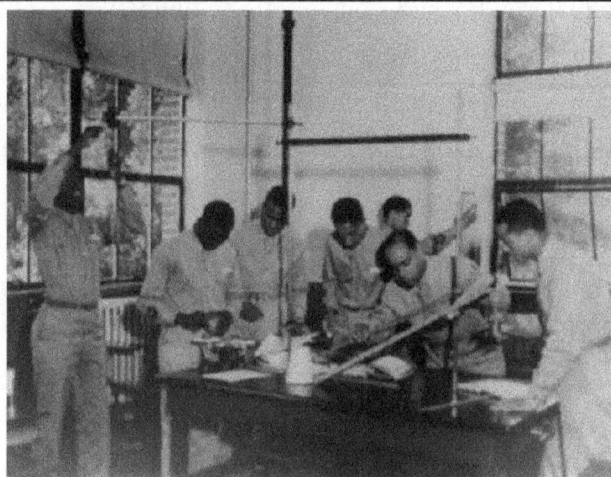

Aviation cadets in the classroom at Tuskegee.
(USAF Photo)

Interpretation of the Selective Training and Service Act of 1940 opened the flying and later aviation cadet programs to African-American applicants, provided they met the program's qualifications. The goals of the CPT program were twofold: to stimulate commercial aviation and to provide a ready pool of pilots in case of war. The program provided students with a ground school familiarizing them with the operation of aircraft, then 18 hours of flight. Because of the dual nature of the program, participants were not required to enlist during the war. The Civilian Training Program was one of many wartime programs that went out of existence in the immediate post-war period.[98]

While the school authorized under PL 18 never materialized, in January 1941 the War Department announced that it would create the 99th Pursuit Squadron, a black flying unit, with a program for training African-American pilots at Tuskegee Institute. Because of its location, the men who trained there became known as the "Tuskegee Airmen."[99]

As with the rest of the statistics for pilot training, there is no detailed break down of aviation cadets who went through the flight program. Since there were so few African-American officers in the military at that time and opportunities for attending Officer Candidate School were limited, one might assume that the great majority of who completed pilot training did so as aviation cadets. Of the 2,433 African-Americans who entered preflight training, a total of 926 completed the advanced phase and received their wings and commissions. The training program mirrored the aviation cadet program modeled on the Randolph Field standard. Unlike other aviation cadets, at least initially, black airmen eliminated from training could not be reassigned to other programs. By late 1943, this disparity was partially corrected and some did enter navigator and bombardier training. A significant problem was the generally low education level among African-Americans. Until the spring of 1944, the stanine scores used to qualify African-Americans for pilot, navigator, and bombardier training were lower than those used to send white applicants into training. By February 1944, however, the scores needed for all aircrew trainees were the same—5 for pilot training, 6 for navigator, and 5 for bombardier.[100]

An indeterminate number of African-American aviation cadets completed navigator and bombardier training and ground duty programs. Unable to provide training for the African-American bombardiers, navigators, and other specialists at Tuskegee, Training Command sent black aviation cadets to other training sites. African-Americans completed navigator training at Hondo Field, Texas, and bombardier training at Midland Field, also in Texas. Black cadets and officers followed the same training regimen as their white counterparts. While relationships

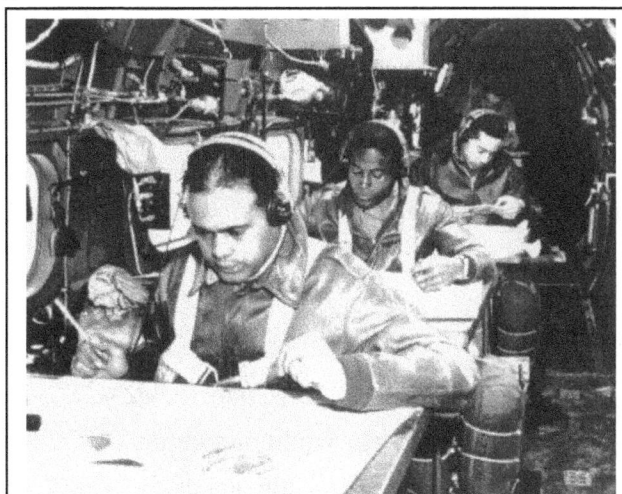

African-American aviation cadets studying to be navigators. **(USAF Photo)**

between the students at Hondo were marked by "distant but peaceful cooperation," a real rapport seemed to develop at Midland. Blacks and whites flew together at the bombardier school and black officers were integrated into the base's officers club. The African-American cadets had a separate club from their white counterparts, but the facilities and amenities were comparable. Altogether, 132 African-Americans completed navigation school and 261 bombardier training. Alan M. Osur in his study, *Blacks in the Army Air Forces during World War II: The Problem of Race Relations*, mentions that there were quotas for seven African-American meteorologists.[101]

Prohibited from serving as military pilots during the First World War, women found limited acceptance into the pilot fraternity during World War II, though in a civil service status instead of as active duty military members. While the women were virtually handpicked for the two flying programs that later became the Women Airforce Service Pilots (WASP) and did not go through the aviation cadet program, they undertook virtually the same training program as their male counterparts.

Nancy Harkness Love and Jackie Cochran, well-known women aviators, were the two leading advocates of allowing women to fly for the Army. In September 1942, the Army Air Forces established the Women's Auxiliary Ferrying Squadron, consisting of 25 women pilots. To free male pilots for war, the women flew airplanes from the factory to airfields in the United States. The AAF established the Women's Flying Training Detachment in October 1942 and used Ellington Field near Houston for the pilot training program. Later, in February 1943, Training Command officials moved the training program to Avenger Field in Sweetwater, Texas. Because of stringent screening processes, of the 25,000 women who applied for training during the war only 1,830 were accepted, and 1,074 won their wings. The women pilots, though not in the military, were generally accorded the same privileges and pay as junior air force officers. Despite the hope that someday the women would receive military commissions, proponents could not get Congress to change the law to grant them military status during the war. Finally, in 1977, the surviving members of the WASP were granted veterans status. In August 1943, the two women's programs were united, forming the Women Airforce Service Pilots.

Women pilots flew all types of aircraft, including B-29 bombers, the largest planes in the arsenal, and proved themselves as capable as men in all areas, except combat, from which they were excluded. In October 1944, as the war entered its final phases, General

Arnold announced the termination of the women's program effective December 20, 1944. At the same time he stated, "You...have shown that you can fly wingtip to wingtip with your brothers. It is on the record that women can fly as well as men." Agitation by male pilots returning home from the war forced women out of military aviation and out

WASP students studying their flight plan for the day.
(USAF Photo)

of many positions in commercial industry.[102]

End of Hostilities
As the war wound down, the AAF significantly reduced the number of students entering its training programs. By the end of 1944, all preflight training had been consolidated into a common course for all aircrew trainees at Randolph Field, and most of the primary contract schools and the more advanced training bases had been closed. By mid-1945, pilot training for white students was restricted to: combat returnee officers, those who had been flight instructors at civilian and educational institutions, and former airline pilots. In June 1945, AAFTC moved its preflight school to Maxwell Field, and by 31 August, the command conducted primary training at only five bases and basic at six. There were four advanced single-engine schools and four two-engine schools; navigator training took place at three schools and bombardier training at five. After the August ceasefire, only those who expressed a desire to serve in the post-war air force continued in training. To slow production, both basic and advanced pilot training were lengthened from 10 weeks to 15 in the fall of 1945. Between May and August 1945, only 848 students entered white pilot training; over 3,100 students who had previously entered training

completed their programs during the same time period. In navigator training, Training Command announced on 25 May 1945 that no more students would enter training at Hondo AAF, and only 118 students would enter the program every two weeks. In September 1945, AAF officials set the annual production rate for navigators at 1,300 and for bombardier-navigators at 152. At the same time, officials lengthened the navigator course from 24 weeks to 30 and the bombardier-navigator course from 40 weeks to 50. As with pilot training, only those who agreed to serve in the post-war air force were allowed to complete training. African-Americans in flying training programs also dropped significantly. During the last three months of 1945, only 46 completed advanced pilot training, none graduated from navigator training, and only 2 completed bombardier training. By the end of 1945, entries into the aviation cadet program were temporarily curtailed.[103]

A room full of Link trainers. An article in the Randolph Field cadet magazine, *Form One*, from February 1943 described this simulator as "a demon, designed to torture Cadets."

The trainer allowed the student to practice:

Straight and level flight, standard-rate turns, climbs and glides with turns, stalls, spins and last but worst, rough-air!

The Link was also used to teach instrument flying when the hoods, visible in this photograph, was lowered.

(USAF Photo)

Chapter IV - Post-World War II Drawdown, Korea, and Vietnam

With the end of World War II, the nation dramatically downscaled its military, as it had following the First World War. In practical terms, the Army and soon-to-be independent United States Air Force had more rated officers than it needed and suspended the aviation cadet program briefly. During the Korean War, the cadet program again ramped up, but senior Air Force officials decided late in the 1950s to establish an officer corps composed of college graduates. Subsequently, the aviation cadet program, which had allowed non-college graduates to gain their commission, ended in the early 1960s. By the time the Air Force needed significant numbers of rated officers for the Vietnam War, the Air Force ROTC program, Officer Training School (OTS),[104] and the USAF Academy provided enough pilot, navigator, and bombardier candidates for training, though some thought the aviation cadet program might continue to serve a useful purpose.

The Immediate Post-War Years

As World War II ended, entries into the aviation cadet program for pilot trainees were suspended. In October 1945, personnel in aircrew training were given three choices with respect to their future with the air force. They could choose to leave the service with an honorable discharge, continue in their training program, or they could revert to a previous enlisted status. By the end of 1945, only a handful of pilot training schools remained active within the command and the preflight training program had ended. The total number of training aircraft on hand dropped from 6,638 in September 1945 to 3,646 in December, while flying hours fell from 218,978 to just 45,114. The training curriculum was extended from 10 to 15 weeks in each of the three phases of pilot training—preflight, primary, and basic.[105] There remained some aviation cadet entries into the bombardier and navigator training programs. In September 1945, all navigator training was consolidated at Ellington Field and all bombardier training at Midland Field, though it took a few weeks to close out training at the other bases. In October 1945, the training curriculum for aviation cadets studying to be navigators and bombardiers was lengthened from 24 to 30 weeks, primarily because the work week was cut from 48 hours to 40 hours.[106] As noted in the previous chapter, there were no ground duty aviation cadet assignments following the end of World War II. Officer Candidate School and the Air Force Reserve Officer Training Corps provided almost all of the non-rated officers needed by the service between 1945 and the end of the aviation cadet era in 1965.[107]

Harold Z. Hewitt and Max J. Christensen, two of the last aviation cadets to complete the World War II-era pilot training program, were caught up in the base closings and program changes in the months following the end of the war. Both enlisted before the war ended and decided to stay in the service. Hewitt and Christensen began basic pilot training at Perrin Field, Texas. When Training Command closed the airfield, officials shipped the remaining aviation cadets in the program to Randolph Field. A large number of Chinese cadets also training at the San Antonio airfield severely limited flying time as the command was retiring training aircraft wholesale. The Americans transferred to Luke Field, Arizona, to complete basic and, subsequently, start advanced training. Luke closed before the men could complete their training program and once again they moved. This time, Hewitt and Christensen had a relatively simple move, just across the Phoenix area to Williams Field. The men finally graduated and received their wings and commissions in October 1946.[108]

Early in 1946, Training Command approved a training moratorium of 7 1/2 weeks to allow its Flying Training Command to retrench and regroup. The command experienced a constant drain on maintenance, supervisory, and instructor personnel as the drawdown continued. By the end of June, only five bases hosted pilot training schools. No new entries were allowed into pilot training programs between 29 January and 26 March, and the graduation dates of U.S. students were delayed by 7 1/2 weeks. Foreign student training received top priority. Between January and June 1946, only 119 students graduated from primary, 75 from basic, and 178 from advanced pilot training.[109] There is no indication how many, if any, of these were aviation cadets, though the number must have been small. At the same time, Training Command began the integration of navigator, bombardier, and radar observer training programs into a single, "aircraft observer, bombardment (AOB)" program. First priority would be the retraining of approximately 2,000 navigators, bombardiers, and radar observers on active duty who needed to qualify for the new triple-specialty. After nearly moving the school to Las Vegas Field, Nevada, at the end of 1946, Air Training Command[110] and Strategic Air Command officials agreed that Mather Field, California, would host the new training program.[111]

In December 1946, HQ AAF announced the resumption of the aviation cadet program for pilot trainees. Initially, unmarried enlisted men between the ages of 18 and 26 1/2 who were of excellent character were eligible to apply. The men also had to possess a high school diploma and be able to pass the classification battery of tests, consisting of 12 paper-and-pencil tests and 6 psychomotor tests. Successful applicants scored at least 7 on the stanine. Finally, the enlisted candidate had to have at least two years of obligated service remaining. In March, 83 of the 480 students who entered the primary phase of pilot training at Randolph Field (the only base at which primary was given) were aviation cadets. In May, having largely exhausted the number of enlisted men willing and qualified to enter the aviation cadet program, the AAF opened applications to civilian personnel. A small group of civilians entered training with the July class. At the same time, officials raised the minimum age to 20 and required applicants to have completed at least two years of college or pass a qualifying education exam. As with enlisted personnel, successful civilian candidates took the classification battery of tests and needed a 7 to enter the program. Most of the civilians who entered the pilot training program had no college experience and qualified based on the education examination. After training, the newly commissioned officers had a three-year active duty obligation; those not receiving a regular commission during that time reverted to an inactive reserve status. Officials abandoned the use of the stanine in October because of the increasing demand for pilot training students. No aviation cadets entered AOB training in 1947.[112]

Aviation cadets gathered around a T-33. Their instructor is describing the upcoming mission.   **(USAF Photo)**

During 1947, the command produced a total of 371 pilots, a goal that the Air Force hoped to increase to 3,000 for its planned peacetime force of 70 combat groups. In 1948 pilot production reached 900, and by April 1949 officials were on track to enter the nearly 600 students needed in each of eight scheduled classes per year to attain the 3,000 pilot training graduates. Because of the expected attrition rates,

ATC officials programmed nearly 4,800 entries into pilot training to produce the desired 3,000 graduates. To meet the expanding program, the Air Force activated Goodfellow, Perrin, and Waco (later redesignated James Connally) bases for basic pilot training and Las Vegas and Enid (now Vance) for advanced in 1947 and 1948.[113]

The pilot training program changed significantly in the years before the Korean War. In September 1947, ATC officials combined what had previously been the separate phases of primary and basic flying training into a single, 8-month-long "integrated basic" course. This training program included a 2-week preflight program during which entering students were introduced to the theory and practice of flying. Advanced training lasted four months under the September 1947 program. During the first half of 1949, the command added 2 weeks to preflight and made both basic and advanced 6 months long. The 4-week preflight better indoctrinated students in the military life before beginning their intensive flying training programs. The 6-month basic and advanced better balanced the training load among the bases. Instructors used the T-6 exclusively in basic and F-51 "Mustangs," B-17 "Flying Fortresses," and B-25 "Mitchells" in advanced training. New to advanced training was the F-80 "Shooting Star," the Air Force's first operational jet fighter. The two-seat trainer version of the F-80 was designated the T-33.[114]

Observer training also changed significantly in the Air Force's "three-headed monster" AOB program. As with pilot production, the Air Force significantly increased the production goal for its AOB program. The annual production goal of 180 in 1947, more than tripled the next year to 564. Through 1948, the command trained only rated officers, but opened the AOB program to aviation cadets in 1949. The Air Force applied the same qualification standards needed for the pilot training program to AOB. The command established a USAF Navigation School at the newly reactivated Ellington AFB[115] on 15 April 1949 to provide training. Aviation cadets and officers without any aircrew training who entered AOB training went to Ellington while officers who had previously been rated went to Mather.[116]

At first, aviation cadets and non-rated officers undertook a 48-week training program at Ellington. Classroom and flying training were the same for all students, except that those who had previously completed military training were excused from classes such as military law. Aviation cadets also completed basic military training in drills and ceremonies and leadership, discipline, and customs much like pilot training cadets. Beginning in November

1949, non-rated officers entered a 38-week program while aviation cadets completed 48 weeks of training. The officers had a shorter training program because they were not required to take the basic military training classes or spend time in drill. Students used the C-47 for navigation missions and B-25s for radar and bombing training.[117]

Korea

With the start of the Korean War in June 1950, aviation cadet recruiting, classifying, and training again became big business for ATC. Unlike the situation at the outbreak of World War II, the number of high-quality applicants for aircrew training was insufficient to meet the needs of the service. From a peacetime production goal of 3,000 pilots, the service ultimately expected Air Training Command to produce 7,200 pilots each year to man its 95 combat wings. Based on a 29 percent attrition rate, ATC needed to get about 10,000 candidates into pilot training. In fiscal year 1951, covering the first year of the Korean War, only 2,119 pilots graduated from training. During that year, lower than needed student loads and higher than expected attrition drove the underproduction. Only 4,556 students entered training, more than half of whom failed to complete it. To meet production demands, the command's

infrastructure grew dramatically. The command reactivated or gained from other major commands 15 bases and, returning to a World War II program, opened 9 contract pilot training schools between 1950 and 1953.[118]

Training officials conducted a comprehensive study, especially concentrating on the 50 percent attrition rate for aviation cadets. Study team members made field trips to the training bases, where they interviewed instructors, supervisors, and students. In addition, questionnaires were sent to former students who had simply resigned from training. The results showed that many aviation cadets felt they had been misled about what to expect while in training. Recruiting advertisements glamorized the prospect of being a pilot without painting a realistic picture of the military discipline, hard work, and living conditions they would experience while in training. The low pay of aviation

TB-25 training aircraft.   **(USAF Photo)**

cadets, $105 per month, was another contributing factor. Especially grating, apparently, was the observation that the pay of the average enlisted trainee was 50 percent higher than that of the aviation cadet. The third major contributing factor to come out of the study was the elaborate class customs and hazing. Most loathsome were customs in the mess halls that kept some cadets at a rigid position of attention while trying to eat. Other customs interfered with evening study. One of the former aviation cadets responding to the questionnaire stated that, "I think customs are a frivolous waste of time and energy and in no way reflect upon one's potentiality as a pilot or officer." Another noted, "They should let you eat and study in peace. My whole class had to eat at the Cadet Club after chow." Despite the frustrations expressed, almost 70 percent of the resignees who responded to the interviews and

Aviation cadets of the 1950s practicing the traditional "square" meal. (USAF Photo)

questionnaires expressed regret at having left the program and 81 percent said they would welcome an opportunity to re-enter the program. Based on the study, ATC established a counseling program at each of the bases to give aviation cadets advice on how to cope with the demands of training program. The command revised the military training portion of the curriculum and consolidated all preflight training at Lackland. The recruiting program was revised to attract more applicants and to more realistically portray what the aviation cadet would likely experience. The issue of pay, however, was beyond the capability of ATC to alleviate.[119]

As another method of combating the high attrition rate, the command changed the aviation cadet qualifying examination to put more emphasis on math, physics, and science and increased the length of the test from 2 hours to 5. Also, in July 1951 the command reinstituted the use of the aircrew classification battery, or stanine, as a method of screening its aircrew candidates. While the Air Force had discontinued the use of the tests as a screening device in 1947, the service continued to administer the tests to aviation cadets entering training during the post-war years for research purposes. Research continued to reflect the World War II experience, the higher the student's stanine score, the better physically and psychologically qualified he was to succeed. If a cut-off score of 9 were set, for example, it would only take 1,097 entries into training to get 1,000 graduates. Setting the stanine qualification at 5, the Air Force would need 1,610 entries into pilot training to produce 1,000 graduates. The trade-off came with the pool of applicants needed to get those who met the stanine requirement. If set at 9, the Air Force would need to screen 13,526 applicants to get the 1,097 qualified to enter training; a stanine of 5 required only 2,403 applicants to get the 1,610 needed to enter training. Instead of spending a lot of money during training on less qualified students, a lesser amount would be spent of screening applicants. The command set its stanine at 6, and officials hoped to cut the attrition rate among aviation cadets to 29 percent.[120]

ATC officials also studied the various qualifications needed to enter pilot training to determine which had the most effect on success in the Air Force. They hoped to eliminate trial-and-error methods of reducing the selection criteria in the future because they were so costly and wasteful. They found that college experience seemed to make little difference. A survey of World War II elimination rates showed that 18 percent of aviation cadets with a high school diploma washed out of aircrew training compared to 15 percent of those with 1-4 years of college. Those with more college did slightly better in the area of

career development; 90 percent of the aviation cadets who came into the service during World War II with at least two years of college had attained the rank of captain or higher compared to 83.7 percent of those with less than two years of college. At the highest level, of the 367 general officers in the Air Force as of October 1951, 22.7 percent had only a high diploma or less education. With respect to age, the study board found that eliminations during World War II were lowest among students 18-19 years old and that the washout rate increased for each higher age group. Special problems accommodating the needs of married applicants and achieving pay equity for noncommissioned officers compared to the relatively few aviation cadets to be drawn from these manpower pools did not seem worth the trouble. During World War II, the stanine had proven to be an effective predictor of success in the aircrew training programs, but there was a trade off. A high stanine meant a higher percentage of the students would graduate, but a greater number needed to be screened. A low stanine increased the number of students who would qualify for training, but increased the operating cost of training programs by increasing the number of washouts. ATC officials also believed that the lower stanine also allowed inferior officer material to join the ranks. By eliminating the 2-year college requirement and lowering the minimum age, the Air Force could maintain a large applicant pool to support a higher stanine and thereby improve the overall quality of aircrew training graduates.[121]

Because pressure to get men into aircrew training continued as the war progressed, USAF officials dropped the requirement for 2 years of college in November 1951 for airmen who had served for at least 18 months, but civilian applicants still needed the college experience. In January 1952, the service requirement for enlisted applicants was removed, and in February the minimum age for all applicants was lowered to 19. The stanine for pilots and AOB candidates followed a downward track to allow more people into training—to 5 in November 1951, and to 3 in April 1952. Another inducement to get volunteers was lowering the 3-year obligation for civilians entering the aviation cadet program to two in December 1951.[122] The changes in the

The Captivair "synthetic" trainer, which simulated flight conditions in the T-33.   (USAF Photo)

qualification criteria nearly achieved their purpose. The pool of applicants increased sufficiently so that by the end of 1953, the command's pilot production neared the 7,200 target, if foreign students were counted in the total.[123]

In November 1952, ATC unveiled a "revitalized" pilot training program in an attempt to lower attrition or, at least, push a greater portion of the washout rate into the earliest phase of the program where the investment in the student was not so great. The centerpiece of the four-phase, revitalized program was the consolidation of all preflight training at Lackland AFB and lengthening preflight from 4 weeks to 12. Previously, each of the primary bases had conducted preflight training. Consolidation of preflight training allowed the command to provide the same basic military training and indoctrination to all aviation cadets. At the same time, the flying training bases could reduce the amount of basic military training they gave to aviation cadets and concentrate on what they did best, flying training. ATC officials also decided to include light plane screening in the preflight program; they hoped that those students prone to wash out for fear of flying or flying deficiencies would do so flying a relatively inexpensive to operate aircraft, such as the Piper Cub, designated the PA-18. The preflight training program for officer trainees was virtually identical to that given to the aviation cadets. The only difference was that student officers completed 162 hours of "officer training" compared to 167 hours for aviation cadets.[124]

As part of the new program, the command adopted an 18-week primary, using the T-6 and T-28, and a 16-week basic training program, using the T-28 for conventional training and the T-33 for students training to be fighter pilots. Crew training, which averaged 12 weeks, was the fourth component of what became a fully integrated program from preflight through specialized crew training. Aviation cadets spent nearly 16 months in training, though they continued to receive their wings and commissions after completing basic. The students of class 53-H were the first to enter the new program. To increase the quality of students going through pilot training, the command increased the qualifying stanine from 3 to 5 in November 1952, then, in July 1953, raised the stanine for non-college graduates entering pilot training to 7.[125]

ATC officials hoped to reduce attrition to 20.4 percent under the revitalized program, significantly lower than the 34.4 percent washout rate experienced when the revitalized program was introduced in 1952. The first classes to graduate under the new curriculum, however, experienced a 34.7 percent attrition rate. This was reduced somewhat, to 32.1 percent, for the January to June 1954 reporting period.

ATC steadily reduced attrition until reaching 26.9 percent during the first six months of 1955. By that time, ATC had raised the planned attrition to 29 percent. Beginning in July 1955, the attrition rate started edging back up.[126]

ATC officials introduced the T-28 "Trojan" during the Korean War. Initial results, however, were less than spectacular. The editor of the 51-H classbook from Reese AFB commented that:

> *To make us appreciate the safety and dependability of the B-25, we were sent aloft in the T-28, at the time the Air Force's newest trainer. Its engine sounded like a garden tractor and could be depended on to go to pieces at any minute. Everyone felt like a test pilot when he flew the machine.[127]*

The command also began planning for a time when specifically designed trainer aircraft would improve the training experience and bring efficiencies to the training program.[128]

Observer training similarly underwent dramatic changes. In September 1951, the command's Flying Training Air Force (FTAF), one of three major ATC subordinate units,[129] implemented a Single Observer Training program and began to phase out the various courses used earlier to qualify previously-rated aircrew personnel. The main reason for the change was the need to provide military training for aviation cadets who were beginning to provide the bulk of the students. The first phase of the program, basic observer training, provided a common curriculum for all students, though three separate tracks were established for: 1) aviation cadets and non-rated officers, 2) navigators and bombardiers coming back onto duty, and 3) pilots who wanted to train as navigators. The basic observer training program for aviation cadets, taught at Ellington and Harlingen AFBs, lasted 24 weeks. The advanced phase, conducted at Ellington and Mather, provided specialized training for specific weapons systems, such as the B-29 and, later, the B-47 bombers. Because the advanced training programs lasted varying lengths, from 12 weeks to 20 weeks, the question of when AOB aviation cadets should be commissioned arose. In May 1952, Air Force officials decided that cadets should be commissioned at the end of 42 weeks of training. Students receiving their commission before completing training would still be required to graduate from the course; if he washed out subsequent to being commissioned, the student's

commission would be revoked and he would be separated from the service. Students whose training program ended prior to commissioning would be required to stay at the training center until receiving a permanent duty assignment. Shortly after the consolidation of pilot preflight training at Lackland, ATC established a 12-week preflight program for AOB cadets at Lackland. The first class of cadets arrived at Lackland in March 1953 and followed much the same curriculum as aviation cadets entering pilot training, though with a greater emphasis on math and physics. AOB production approximated Air Force requirements. In the last half of 1953, for example, the command graduated 1,457 aviation

T-28s flying in formation.   (USAF Photo)

cadets and non-rated officers from the basic observer course against a goal of 1,474.[130]

The principle AOB training aircraft at Ellington AFB were the familiar TC-47 and TB-25. The command used the TC-47 for all types of instruction except radar navigation, which was provided in the TB-25. At Harlingen, students trained in the TC-45 and the newly introduced T-29. Mather students flew the T-29, TB-25, and TB-50. The T-29s, the most versatile of the aircraft, were used for both conventional and radar navigation and for simulated radar bombing missions and acquired the nickname, the "Flying Classroom."[131] ATC eventually adopted the T-29 as its standard AOB trainer aircraft.

During the Korean War, the command gained major responsibilities in managing aviation cadet recruiting and screening. While advertising and policy making remained an Air Staff function, officials delegated responsibility for final processing, qualification, and selection of aviation cadet applicants to ATC. To implement the changes, in November 1951 ATC tasked its FTAF with program management, though ATC continued to run the initial classification testing units and examining boards. To help recruit applicants for the cadet program, FTAF established two experimental recruiting teams, one located at Scott AFB and one at Chanute, in January 1952. Team members visited colleges and universities in the states of Illinois, Indiana, and Missouri to publicize the cadet program and function as aviation cadet examining boards. Based on their success, FTAF added 10 more teams in April and May 1952 and eventually fielded 72 teams in early months of 1953. The host bases at which the teams were located provided logistical support. In addition to visiting educational institutions, the teams contacted civic organizations and set up exhibits at events such as state and county fairs. The command also established the 3500th Personnel Processing Squadron (Aviation Cadet) in April 1952 to

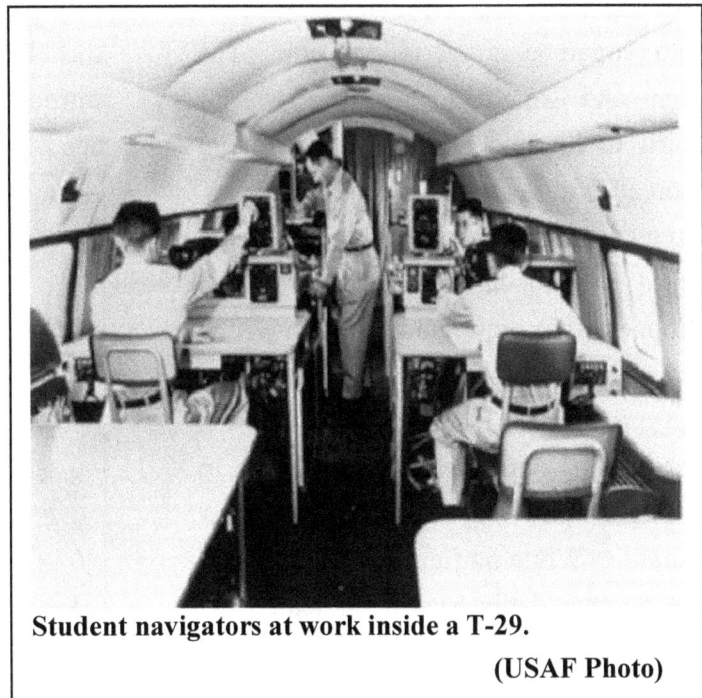

Student navigators at work inside a T-29.

(USAF Photo)

provide operational control over the recruiting teams, cadet examining boards, and the aircrew classification testing units. This unit was collocated with FTAF headquarters in downtown Waco, Texas.[132]

An ATC study of student motivation conducted in August 1952, led officials to believe that the high attrition rate in the pilot training program, at least in part, could be lowered through the media used in the aviation cadet recruiting program. Print material emphasized the opportunity for service instead of the chance to learn a trade for personal gain. To meet the increasing need for fighter pilots, recruiting material focused on single-seat, jet training. The widespread use of acrobatic teams such as ATC's Acro-Jets and USAFE's Sky Blazers, the forerunners of the Thunderbirds, came into vogue as a recruiting tool. The command sought to instill a warrior spirit in its student pilots and brought Korean War veterans into the classroom. To qualify more men for fighter training, on 3 December 1952 the Air Force dropped the requirement that no one over 72 inches (or 6 feet) tall could enter the program. Instead, so long as a person was not taller than 38 inches in sitting height, they could enter fighter training. While no figures were found in the command histories to indicate how many students over 72 inches tall met the 38 inch sitting height criterion, the

July-December 1952 FTAF history identified 414 students over 72 inches tall currently in other pilot training programs who had desired to enter fighter training.[133]

Because the motivation study was called "Project Tiger," tiger roars became part of the aviation cadet ritual at preflight. When asked by an upperclassman what was on the menu for supper, the underclassman was required to answer, "Raw meat and tiger juice, sir." A pamphlet published in September 1954 highlighted motivation as the key to success and presented hazing in a positive light. Not to be confused with the light-hearted horseplay that might be found in a college fraternity, hazing strengthened the cadet's emotional balance and gave him a chance to practice self-discipline. The pamphlet contained photographs of the latest aircraft in the fleet, an introduction written by a Korean War ace, and a description of the various

Acro-Jets flying in close formation.
(USAF Photo)

phases of training. By the time the student reached advanced training, social opportunities, illustrated by photographs of aviation cadets both at a swimming pool surrounded by young women and at a dance, could be enjoyed. The last page of text told the cadet-to-be that he would serve with pride and a sense of exhilaration. As an Air Force pilot, it would be possible to "serve a tour of duty in the Far East, Europe, Africa, or vacation on the French Riviera, stop off in the Philippines."[134]

The Pre-Aviation Cadet Program

By early 1953, nearly 1,800 enlisted men had qualified for the aviation cadet program but school quotas were not available; further, ATC officials believed that between 1,200 and 2,000 would be in an awaiting training status for the foreseeable future. To keep these men motivated and interested in the cadet program, officials developed the pre-aviation cadet program. Nine ATC bases created pre-cadet detachments to administer the program. The command published a standardized, 2-week curriculum in January 1953 that provided 60 hours of academic training. ATC officials created a distinctive insignia for the pre-cadets—a bright red loop approximately 1-inch wide worn on the shoulder straps of their uniforms. In addition to the pre-cadet class work, the airmen were assigned duties in aircraft maintenance or operations units for that

would prove valuable during their pilot or AOB training. As openings became available, those who had completed the pre-cadet program entered the 12-week preflight and pre-AOB training programs at Lackland.[135]

With the end of the Korean War in July 1953, the pre-aviation cadet program was discontinued at six of the nine bases by the end of the year. Only the programs at Goodfellow, James Connally, and Reese continued. In November 1953, the command decided to process all of the enlisted personnel interested in the cadet program through its indoctrination centers at Lackland AFB and Sampson AFB, New York. Those entering the pilot training program through Lackland reported to Goodfellow, those qualified through Sampson went to Reese. All enlisted personnel interested and qualified for the AOB program went to James Connally.[136]

In February 1955, ATC officials revamped and expanded the pre-cadet training program from 60 hours to 90 hours of instruction. The main addition to the curriculum was a 25-hour block of mathematics. Base officials were allowed up to 9 weeks to provide the 90 hours of training. At the same time, ATC re-established pre-cadet programs at Edward Gary AFB, Texas, Vance AFB, Oklahoma, and Williams AFB. A commitment to enter more student officers, especially AFROTC graduates, than enlisted men into the aviation cadet program brought an end to the pre-cadet program in May 1956.[137]

Demise of the Aviation Cadet Program

After ramping up pilot recruiting and production to meet the requirements of the Korean War, the Air Staff informed ATC in September 1954 that, beginning in July 1956, the production of 4,800 pilots per year would be sufficient. The production goals followed a steadily downward trend through the rest of the decade. The command expected to provide students higher quality training as the wartime push was over by reducing class size and increasing flying time.[138]

The post-war period saw some changes in the qualifications needed for entry into the aviation cadet program. At war's end, an applicant for pilot training with two years of college needed to score a 5 on the stanine, while

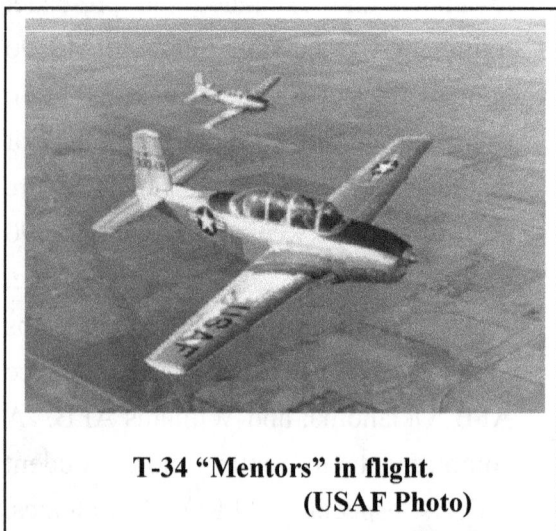

T-34 "Mentors" in flight.
(USAF Photo)

high school graduates needed a 7. To enter the navigator training program, applicants needed only a high school diploma and a stanine of 5. The stanine for both programs was lowered to 4 for applicants possessing a college degree in October 1959. At the same time, officials increased the stanine score to 8 for high school graduates coming into the service from civilian life to enter pilot training; enlisted men with a high school diploma who desired to train as pilots needed a 5. The October 1959 guidelines also specified that OTS and ROTC graduates who wanted to be pilots or navigators needed a stanine of 4 to enter both pilot training and navigator training. Beginning in December 1961, the USAF once again required navigator cadet applicants to have at least two years of college.[139]

In March 1954, the Secretary of Defense gave the USAF operational control of Air Force recruiting programs; in turn, the Air Staff redelegated the responsibility to ATC. With a unit already devoted to recruiting civilians for its aviation cadet program, ATC expanded the scope of its personnel processing force to include all airmen. The newly activated 3500th Recruiting Wing was relocated from Waco to Wright-Patterson AFB, Ohio, to be closer to the more heavily populated areas of the U.S. With the creation of the 3500th Recruiting Wing in 1954 and the assignment of all Air Force recruiting activities to ATC, the aviation cadet recruiting program transferred from FTAF to the 3500th. In 1959, ATC inactivated the recruiting wing and in its place activated the USAF Recruiting Service.[140]

Open-bay barracks at the Lackland preflight school.
(USAF Photo)

In May 1954, with Congressional prompting, the Air Force made a commitment to bring more AFROTC students into pilot training. On March 1955, for every 88 aviation cadets graduating from pilot training, there were only 12 AFROTC student officers. By 30 June, the numbers were a 78/22 balance, but the proportion of students entering training was split 50/50. ATC planned to reach a 35/65 mix of aviation cadets to AFROTC graduates in the near term.[141]

The Soviet Union's successful launch of Sputnik, the world's first man-made satellite, in October 1957, reinforced the perceived need for a more scientifically- and technologically-oriented Air Force as well as a more

educated officer corps. The increasing complexity of weapons systems— epitomized by an air fleet dominated by jet aircraft and by advances in radar— was already changing the skills and training needed by the force. At the time of the Sputnik launch, the USAF's two largest warfighting commands, the Strategic Air Command (SAC) and the Tactical Air Command (TAC), had an officer corps overwhelmingly consisting of non-college graduates. Only about 31 percent of SAC and TAC officers held bachelors degrees and about 1.5 percent had masters degrees.[142]

A fundamental change in the students sent to training involved the mix of cadets and student officers. In 1952, 70 of every 100 students in rated officer training programs were aviation cadets; by the last years of the cadet program, 70 percent of students in aircrew training programs had already received their commissions. This dramatic shift followed the creation of the Air Force Academy in 1955 and the OTS in 1959 and an emphasis on the development of an officer corps consisting of college graduates. In January 1961, the Air Staff directed ATC to recruit more engineering and scientific students for OTS. The goal was to bolster the education levels of the communications-electronics, guided missile, and engineering officers. After 1961, led by General Curtis LeMay, the Air Force Chief of Staff, the service looked forward to a future when at least 95 percent of its officers had college degrees. Table I shows the source of commissioning for officers during the last years of the aviation cadet program.[143]

ATC officials carefully watched the performance of USAF Academy graduates. A study completed of the 1959 USAFA graduates who had entered pilot training found that of the 186 academy grads who entered training, only 12, or 6.5 percent, were eliminated. Among other Air Force commissioning sources, attrition rates were significantly higher. Over half of the OCS graduates in pilot training had washed out, while 42.6 percent of the aviation cadets had attritted. ROTC students who had received flight indoctrination training did relatively well, with only a 15.9 percent washout rate, compared to 37.9 percent of ROTC students who had not gone through flight indoctrination. The study's author commented that the OCS washout rate was abnormally high. Apparently, no OTS students completed pilot training during the study period. Among the class of 1960, 15.5 percent of those who entered pilot training washed out. OCS students continued to be eliminated at a high rate, 40.6 percent; ROTC students with introductory flight training washed out at a 17.6 percent rate; and 33.6 percent of those without such training were eliminated. Aviation cadet attrition increased to 50 percent, though this was artificially high as it represented only 2 of 4 students at the very end of the cadet program.[144]

Pilot and AOB training programs continued to evolve. In 1960, ATC closed its contract primary schools and instituted a consolidated undergraduate pilot training (UPT) program consisting

	1960	1961	1962	1963	1964	1965
Table I. USAF Officer Procurement, Fiscal Years 1960-1965						
USAFA/West Point	268	270	290	489	495	505
AFROTC	3,672	3,270	3,258	3,385	3,962	4,509
OTS	322	525	2,268	5,375	4,438	3,582
Aviation Cadet	1,598	2,149	522	492	211	159
OCS*	424	396	259	-	-	-

*OCS was phased out with the establishment of OTS in 1959.

Source: Vance O. Mitchell, Air Force Officers Personnel Policy Development 1944-1974 (Washington DC: Air Force History and Museums Program, 1996), p. 224.

of three phases—preflight, primary, and basic. All of the phases were taught at the same base and only military instructors were used in training. After Class 62B graduated on 21 December 1960, the command cancelled the next three classes to allow the transfer of equipment, aircraft, and personnel from the primary contract schools to the seven UPT bases: Craig, Laredo, Moody, Reese, Vance, Webb, and Williams. Class 62-F, which entered training on 13 March 1961, was the first to begin UPT. By this time, the production goal for UPT was 1,100 USAF pilots. The 55-week curriculum included 132 flying hours in the T-37 and 130 hours in the T-33.[145]

A major accomplishment in pilot training was the implementation of the pilot trainer aircraft master plan developed in 1952. The T-34 "Mentor" replaced the PA-18 as the light plane screener beginning in May 1954, and the command retired the T-6 in favor of the T-28 as its primary training aircraft. In turn, the T-28 was phased out as the command developed an all-jet training program. The unnamed "TX" that ATC officials looked forward to eventually adding to primary became the T-37 "Tweet." First introduced into the training inventory in January 1958, the T-37 was being phased out of the command inventory as this study was being written. The T-33 remained the workhorse in basic and then UPT until

the end of the aviation cadet era. As the last pilot cadets graduated, the "TZ" of the 1952 plan, the supersonic T-38 "Talon," was undergoing tests at Randolph AFB.[146]

While the command was restructuring its pilot training program, ATC was also making extensive revisions to its navigator training program. Unlike in earlier years, the navigator program was beset by attrition problems in the late 1950s and early 1960s. The main problem seemed to be the 5-year active duty obligation that went into effect in December 1957. After missing a training goal of 1,700 navigators for 1959, the Air Force lowered the active duty requirement for its aviation cadets to 2 years in December 1959. A change in the curriculum also seemed to make a positive effect on production. Beginning with Classes 61-02 at James Connally and 61-03 at Harlingen, ATC established a new "primary-basic" training program, with preflight, primary, and basic phases integrated into a single program. The command issued the syllabus for the 175-day course in March 1960. Attrition in the combined course was programmed at 30 percent, but the command actually experienced a 19.9 percent attrition between July and December 1960. Despite missing the single-year goal for 1959, ATC met the three-year goal of 5,100 navigators between the years of 1959 and 1961. Beginning in 1961, the command started referring to this navigator training program as "undergraduate navigator training (UNT)." Instructors at James Connally and Harlingen AFB conducted UNT; Mather AFB hosted advanced training. Harlingen was identified for closure in 1963 and the command consolidated UNT at Connally.[147]

Phase Out of the Program

In 1957, the Air Force introduced the Air Force Officer Qualifying Test (AFOQT) as a replacement for the Aviation Cadet Qualifying Exam. The emphasis in the new test was on choosing men for the aviation cadet program who would make good officers instead of technically proficient pilots and navigators. The AFOQT, like the old classification battery, was actually a series of tests that were combined into standardized composite scores.[148]

By 1958, the attrition rate in the aviation cadet programs was running higher than expected. ATC raised the planned washout rate among pilot candidates at preflight from 15 percent to 25 and for navigation students in preflight from 15 percent to 30 in October 1958. The command offered two solutions to meeting the expected shortfall in production–live with the shortage until production increased or lower the stanines, then 8 to enter pilot

training and 5 for navigator training. ATC recommended, and HQ USAF accepted, that the service live with the shortage in production. In January 1959, the Air Staff relieved some of the pressure on the pilot program by reducing the planned production of pilots in all categories from 2,700 in fiscal year 1961 to 2,300.[149]

One of the newly introduced T-37 "Tweets."
(USAF

Problems recruiting aviation cadets for the pilot program continued in 1959 and substantial problems showed up in recruiting for the navigation cadet program. Between July and December, ATC hoped to recruit 1,444 pilot cadets but only accessioned 848. For the navigator program, the command set a goal of 1,950 but only recruited 1,179 cadets. The immediate response was to recruit for the aviation cadet program more vigorously, but on 23 December the Air Staff instructed ATC to stop recruiting aviation cadets for pilot training. The final decision to end the program may not have been made at that time, but the Flying Training Program Guidance (PG) issued on 10 February contained no line item for aviation cadets, except for candidates from the Air National Guard. In the interim between the December message and the February PG, HQ USAF had authorized ATC to close its preflight schools at Lackland. There was no need to re-create a pilot preflight school at any of the primary bases because the last class of aviation cadets had already begun training. Six-week navigator preflight schools were to be established at Connally and Harlingen.[150]

The last pilot trained as an aviation cadet, William F. Wesson, completed training on 11 October 1961. He had originally entered training in December 1959 as a member of class 61-F. Wesson broke his back and hip in an accident while in training at Webb AFB when he was forced to eject from his T-33 and lost six months. Initially deemed physically unqualified for flight duty, Wesson fought his dismissal and wound up being reinstated. In June 1961 he arrived back at Webb for training, the only aviation cadet at the base. In fact, ATC created a separate class designator, 62-B-2, for Wesson, since he would not be able to finish his training program in time to coincide with the last classes at other bases.[151]

As a higher percentage of officer students entered navigator training beginning in 1959, it became increasingly obvious that it cost the command significantly more to train aviation cadets. In 1959, 46.3 percent of aviation cadets in the navigator program washed out compared to 19.5 percent of the student officers in training. The situation was even worse in 1960 when 54.4 percent of cadets were eliminated while only 9.5 percent of student officers failed to complete the program. Over the next three years, aviation cadet attrition improved, averaging about 33 percent, but the student officer elimination rate over the same period averaged about 8 percent. Sometime in the summer of 1963 the decision was made to end the navigator program for aviation cadets. A HQ ATC letter dated 26 July discussing OTS procurement stated, "It is imperative that the navigator goal be met since the Aviation Cadet Program will be discontinued after Class 65-15 enters training 5 March." An 18 October message from HQ ATC to the USAF Recruiting Service suspended procurement of navigator aviation cadets effective 1 November. In March 1964, the last navigator cadets entered training.[152]

In a ceremony held at James Connally AFB on 3 March 1965, Steven V. Harper, the last of the aviation cadets, graduated from the USAF navigator training program. Maj Gen Benjamin D. Foulois, one of the first air force pilots, was the keynote speaker and presented wings to the graduates of class 65-15.

Lt Gen Russell C. Davis, who retired in November 2002, is believed to have been the last aviation cadet on duty.

An *Airman* magazine article commented about the poignancy of the event, even though the moment seemingly had little historical impact on the graduates. Col Jasper N. Bell, commander of the 3565th Navigator Training wing at Connally,

had been an aviation cadet in the first class to enter training at Randolph Field in 1931; the training group commander, Col Charles B. Linganmfelter, also trained as an aviation cadet. Harper commented that the first two weeks of his training program had been the hardest, though every facet of daily living throughout the program had been subject to regimentation. Initially, he thought that "entering cadets was the worst mistake I had made in my life," but he decided to persevere.[153]

Conclusion

And so the aviation cadet program, once the major source of rated officers for the Air Force, passed into history after nearly 50 years of existence. While hard data on the number of flying and training cadets is virtually impossible to obtain, it is probably safe to say that over 500,000 men went through the various aircrew and ground duty officer training programs. From the First World War, when the United States Army patterned its flying cadet training program on those of its allies, Air Service and Air Corps officials during the interwar era of the 1920s and 1930s made significant advances in developing a science of pilot training. From this foundation, the World War II cadet programs blossomed into a potent force development program. The use of a classification battery of tests, with its assignment of a stanine score, revolutionized the way men were assigned to aircrew training programs. During the war, it might be argued that navigation and bombardier training gained nearly equal importance with pilot training. The changing technology of the post-World War II era, with the introduction of jet aircraft and increasingly sophisticated technologies, necessarily changed the aviation cadet training programs. Finally, it was the new technologies that brought about the end of the aviation cadet program. Air Force leaders increasingly came to believe that the officer corps needed to be college-educated. With the creation of the Air Force Academy and the establishment of OTS, the cadet program, a venue for men with only a high school education to become air force officers, no longer fit the philosophical needs of the service.

At various times over the next several years after the program ended, many questioned the need for pilots to be college graduates and whether the USAF could benefit from a revival of the aviation cadet program. The Rand Corporation, for example, convened a panel of experts in February 1970 to discuss pilot training in the Air Force. The symposium, instigated by the Air Force under its sponsored-research program with Rand, included representatives from ATC, the Air Staff, and Air Force Systems Command, as well as officers involved with officer selection and training with Britain's Royal Air Force, the Canadian Armed Forces, U.S. Army and Navy aviation training, and American Airlines. The symposium acknowledged that:

The high cost of pilot training, combined with tightening constraints on military expenditures, the rapid growth of technical and scientific knowledge, the new technology, and the imminent need for a refurbishing of aging training equipment, gives new urgency to a reexamination of recurrent issues of pilot training.[154]

Education was the aspect of the discussion that touched most directly on the defunct aviation cadet program. While current USAF policy made a college degree a prerequisite for pilot training, most panelists believed that a college education was not essential to pilot performance and that it would be more advantageous to hire high school graduates. Symposium participants pointed to the experience from World War II and the then-current foreign pilot training programs within ATC. High school graduates from foreign nations training to be pilots had a 20 percent attrition rate. Education seemed to have a more direct bearing on officership. The consensus reached by the symposium participants was that high school graduates should be recruited, trained as pilots, and then sent to college. In this way, the service would have young men at the peak of their physical skills in the cockpit, then they could train the most apt to be career officers. The Air Force took no action on the symposium report.[155]

Looking back on the history of the program in an article published in the July-August 1965 issue of *Air University Review*, Capt Maurice G. Stack wrote about his personal experiences as well.

For most cadets, the cadet program was their first time away from home, the first time they had to rely on their own abilities. Many had never been in an airplane, and many in the early years of the program may not have ever seen an airplane. Stack explained that "cadet life bred confidence…the self-assurance of a man who knows his job, has faced and overcome dangers, and has gained the poise which comes from achievement." As for the demise of the program, Stack realized that the space-age Air Force required all rated officers to hold college degrees; therefore, the aviation cadet program had become obsolete.[156]

One measure of the success of the aviation cadet program is reflected in the sheer number of pilots, navigators, and ground duty officers trained during World War II. Without a program to train men with only high school diplomas for service in the nation's air force, the impact of air power on the war would have been significantly less. In turn, many of these men, and their predecessors, guided the USAF through the Korean War and Vietnam. As shown in the Appendix, in 1950 about 30 percent of general officers were former aviation cadets. By 1970, 25 years after the end of World War II, over half of USAF general officers then on duty were cadet program graduates.[157]

Appendix

Sources of Commissions
Air Force Rated General Officers
(in percents)

	Aviation Cadets	West Point	ROTC	Other
1950				
General	0.9	3.7	-	-
Lt General	2.8	-	-	-
Maj General	9.2	18.3	0.9	2.8
Brig General	15.6	35.8	-	10.1
1960				
General	1.1	4.3	-	-
Lt General	1.1	7.5	1.1	1.1
Maj General	21.5	17.2	1.1	4.3
Brig General	18.3	14.0	4.3	3.2
1970				
General	3.1	3.1	-	-
Lt General	6.1	2.0	3.1	1.0
Maj General	21.4	10.2	2.0	1.0
Brig General	21.4	20.4	1.0	4.1

SOURCE: Vance O. Mitchell, Air Force Officers Personnel Policy Development 1944-1974 (Washington DC: Air Force History and Museums Program, 1996), p. 301.

NOTES

CHAPTER I

1 Juliette A. Hennessy, <u>The United States Air Arm April 1861 to April 1917</u> (Washington DC: Office of Air Force History, 1985), pp. 1-27, passim; see p. 217 for War Department memorandum establishing the Aeronautical Division and p. 225 for an abbreviated version of the specifications for a "heavier-than-air flying machine."

2 Ibid., pp. 15-19.

3 See Volume 1 of <u>The Papers</u>, pp. 493-495, for the Wrights attempts to interest the U.S. government in their airplane and Hennessy, <u>The United States Army Air Arm</u>, pp. 25-34.

4 Hennessey, <u>The United States Army Air Arm</u>, pp. 58-59 and 227; see pp. 236-250 for lists of Army aviators who learned to fly prior to 6 April 1917, when the United States entered World War I.

5 See Hennessy, <u>The United States Army Air Arm</u>, pp. 39-106, passim, and Rebecca Hancock Cameron, <u>Training to Fly: Military Flight Training 1907-1945</u> (Washington DC: Air Force History and Museums Program, 1999), pp. 21-97, passim, for an overview of the early schools.

6 See Charles deForest Chandler and Frank P. Lahm, <u>How Our Army Grew Wings: Airmen and Aircraft before 1914</u> (New York: Ronald Press, 1943), pp. 310-311 for General Order Number 39. See Bruce D. Callender, "They Wanted Wings," <u>Air Force Magazine</u>, January 1991, pp. 80-83 and books such as Anthony Aldebol, <u>Decorations, Medals, Ribbons, Badges and Insignia of the United States Air Force</u> (Fountain Inn SC: Medals of America, 1999); Jon A. Maguire, <u>Silver Wings, Pinks & Greens: Uniforms, Wings & Insignia of USAAF Airmen in World War II</u> (Atglen PA: Schiffer Military/ Aviation History, 1994); and Evans E. Kerrigan, <u>American Badges and Insignia</u> (New York: Viking Press, 1967) for the evolution of wings. See Hennessy, <u>The United States Army Air Arm</u>, pp. 58-59, 234, and p. 229 for a list of the 24 original Military Aviators. The original source citation for the 1916 law is <u>United States Statutes at Large</u>, Vol. 39, Part 1 (Washington DC: GPO, 1917), pp. 166-176.

[7] Hennessey, The United States Army Air Arm, pp. 110-111; see also pp. 233-235 for the text of H.R. 5304. The original source citation for the 1914 law is United States Statutes at Large, Vol. 38, Part 1 (Washington DC: GPO, 1915), pp. 514-516.

[8] Hennessey, The United States Army Air Arm, p. 128.

[9] "Report of Operations of the First Aero Squadron, Signal Corps, with Punitive Expedition, U.S.A. for Period March 15 to August 15, 1916," available from the Air Force Historical Research Agency, Maxwell AFB, Alabama. The "Report of Operations" was also published in Frank Tompkins, Chasing Villa (Harrisburg PA: 1934), pp. 236-245. See also Roger G. Miller, A Preliminary to War: The 1st Aero Squadron and the Mexican Punitive Expedition of 1916 (Washington DC: Air Force History and Museums Program, 2003); Hennessey, The United States Army Air Arm, pp. 167-176; and A. Timothy Warnock, Air Force Combat Medals, Streamers, and Campaigns (Washington DC: Office of Air Force History, 1990), pp. 10-12.

[10] Hennessey, The United States Army Air Arm, pp. 141-163; Cameron, Training to Fly, pp. 24-65.

[11] Hennessey, The United States Army Air Arm, p. 163.

[12] Maurer Maurer, The U.S. Air Service in World War I (Maxwell AFB AL: Albert F. Simpson Historical Research Center, 1978), a four-volume collection of documents, is the best source for World War I history.

[13] United States Statutes at Large, Vol. 40, Part 1 (Washington DC: GPO, 191), p. 42 for the passages in the 1917 law and p. 849 for the 1918 legislation.

[14] Chapter 10 of the "Final Report of the Chief of Air Service, AEF," included in Vol. I of Maurer, The U.S. Air Service in World War I, describes the overseas Air Service training program. See Chapter 1 of Vol. I of the History, AAF Flying Training Command and Its Predecessors, 1 January 1939 to 7 July 1943, available at the AFHRA and at the AETC History Office, for a history of the stateside training program. Army Air Forces Flying Training Command historians writing during World War II commented that the statistics of those who entered training during World War I were "somewhat misleading and contradictory, and in no case, apparently, authoritative." See also studies

such as Douglas A. Galipeau, "Issodun: The Making of America's First Eagles," Air Command and Staff College student paper, Maxwell AFB AL, March 1997, available at http://www.au.af.mil/au/database/projects/ay1997/acsc/97-0602C.pdf accessed 10 December 2003. Cameron, Training to Fly, covers the World War I flying training programs at pp. 107-199.

[15] AAFTC History, 1 January 1939-7 July 1943, Vol. I, pp. 4-8; Cameron, Training to Fly, pp. 121-145. See also Millie Glasebrook, ed., American Aviators in the Great War 1914-1918 (Carson City NV: Glasebrook Foundation, 1984), pp. 133-134, 136-139, 141, and 145.

[16] Cameron, Training to Fly, pp. 131-132.

[17] Ibid.; AAFTC History, 1 January 1939-7 July 1943, Vol. I, p. 6.

[18] Kilner's report is in Maurer, U.S. Air Service in World War I, Vol. I, pp. 319-332; "the wide advertisement…" is at p. 324.

[19] Cameron, Training to Fly, p. 121.

[20] Ibid., pp. 127-128; Glasebrook, American Aviators in the Great War, pp. 133-134; AAFTC History, 1 January 1939-7 July 1943, Vol. I, p. 6.

[21] Edward L. Thorndike, "The Selection of Military Aviators: Mental and Moral Qualities," U.S. Air Service, June 1919, pp. 14-16; see also the August 1919 and the January 1920 issues of this magazine. The original study, "Report of the Investigations of Physiological and Psychological Tests of Aptitude for Flying," dated 5 July 1918, is available from the AFHRA under call no. 141.24-15, Vol. 2; the IRIS no. is 114256.

[22] V. A. C. Hermon, "Air Service Tests of Aptitude for Flying," was published in The Journal of Applied Psychology, 1919, but was obtained for this study from the AFHRA under call no. 141.24-15, Vol. 2, as part of IRIS no. 114256.

[23] See, for example, Melvin S. Majesty, "New Centralized Selection System for Air Force Pilots" (Lackland AFB: School of Military Sciences, Officer, November 1973); Joseph L. Weeks and Warren E. Zelenski, "Entry to USAF Undergraduate Flying Training" (Mesa

AZ: USAF Research Laboratory, August 1998); and Joseph L. Weeks, "USAF Pilot Selection" (Mesa AZ: USAF Research Laboratory, April 2000).

[24] History, School of Military Aeronautics, University of Texas-Austin, pp. 142-145; available at the Center for American Studies, UT-Austin, in the T. S. Painter Collection.

[25] History, SMA UT, pp. 58, 76, 127; "who were morally" from p. 127. At some point, all cadets began wearing the white hat band.

[26] John C. Higdon, "The School of Military Aeronautics," The Alcalde, December 1917, pp. 130-132.

[27] Clarence D. Chamberlin, Record Flights (Philadelphia: Dorrance and Company, 1928), pp. 192-206.

[28] History, SMA UT, pp. 48, 86-88; Merrill Jensen, The Aviation Cadet Ground Duty Program: Policy, Procurement, and Assignment, Army Air Forces Historical Studies: No. 21 (Washington DC: Historical Division, November 1944), pp. 2-3.

CHAPTER II

[29] Statistical data from Office of Statistical Control, "Army Air Forces Statistical Digest World War II" (Washington DC: USAAF, December 1945), pp. 15 for personnel assigned and p. 297 for budgets; Cameron, Training to Fly, p. 132.

[30] The book to read about the interwar air force is Maurer Maurer, Aviation in the U.S. Army 1919-1939 (Washington DC: Office of Air Force History, 1987); see chapters IV and XII, especially, for information about the flying cadet and training programs. See also A History of Military Aviation in San Antonio (Randolph AFB: AETC History Office, September 2000; rev. ed.), p. 74; History, AAFTC, 1 January 1939-7 July 1943 (Fort Worth: AAFTC, 1 March 1945), pp. 17-19 and 26; Flying Cadet Battalion, Roster of Students of the Air Corps Primary Flying Schools (San Antonio: Naylor Printing Company, [1933]), p. 43. The detailed number of personnel assigned and authorized varies among sources.

[31] History, AAFTC, 1 July 1939-7 July 1943, Vol. I, pp. 19-25; "regular and systematic training" from p. 19; Air Corps, Aviation as a Career (Washington DC: GPO, 1928), p. 3; Maurer, Aviation in the US. Army, pp. 53-54. See United States Statutes at Large, Vol. 41, Part 1 (Washington DC: GPO, 1921), p. 109 for the creation of the post-World War I flying cadet program.

[32] History, AAFTC, 1 July 1939-7 July 1943, Vol. I, pp. 19-25; History of Military Aviation in San Antonio, pp. 48-50.

[33] The lighter-than-air branch of the Air Corps was never very large, though the air service commissioned the construction of several airships and balloons during the interwar years. In early 1933, for example, 20 officers and 650 enlisted men constituted the LTA branch. At that time, the Air Corps had four airships, two observation balloons in service and four in storage, and about a dozen free balloons used for training and racing. The Balloon and Airship School closed in the late 1920s. See Maurer, Aviation in the U.S. Army, pp. 372-373.

[34] "Training Schedule," Kelly Field News Letter, September 11, 1920, pp. 1-2.

[35] Ibid., pp. 2-4; "Personnel," Kelly Field News Letter, March 26, 1921, p. 2.

[36] Charles A. Lindbergh, We (New York: G. P. Putnam's Sons, 1927), p. 105.

[37] Lindbergh, We, pp. 104-121.

[38] Benzine was a commonly used cleaning agent, so "Benzine boards" came into Army slang to refer to review boards such as this, with their potential to washout students, or to cleanse the program of those not able to handle the curriculum.

[39] Lindbergh, We, pp. 113-116; "with the washing out" from p. 115.

[40] Lindbergh, We, pp. 126-152.

[41] Memo, Academic Department, Air Corps Advanced Flying School to Commandant, Air Corps Advanced Flying School, "Study on Air Corps Training," 26 July 1926, available from AFHRA as Call No. 248.121-7, IRIS No. 1575511; Air Corps Training

Center, [Diary, 1926-1930], available from AFHRA as Call No. 248.211-56c, IRIS No. 159902.

[42] Air Corps, <u>Aviation as a Career</u>, pp. 2-12; J. E. Chaney, "The Selection and Training of Military Airplane Pilots," <u>U.S. Air Services</u>, March 1928, pp. 21-22; <u>History of Military Aviation in San Antonio</u>, p. 21. Information about those who took the education tests is from <u>Initial Selection of Candidates for Pilot, Bombardier, and Navigator Training, AAF Historical Studies: No. 2</u> (Washington DC: Historical Division, November 1943), p. 10.

[43] <u>Initial Selection of Candidates for Pilot, Bombardier, and Navigator Training</u> contains a table at page 10 showing the results of the education exam between 1923 and 1942. In 1928, only 5 of the 363 men taking the test passed. This was the lowest rate recorded in the study.

[44] Memo, W. H. Hardy, Adjutant [Advanced Flying School] to Chief of the Air Corps, "History of Kelly Field," December 13, 1935.

[45] <u>History of Randolph Field</u>, Vol. I, (Randolph Field: Central Instructors School Office of the Historical Editor, 1944), available from the AETC History Office, hereinafter <u>History of Randolph Field</u>, pp. 119-120.

[46] Brown, <u>Where Eagles Land</u>, pp. 79-80; "Securing Randolph Field for S. A. Called Greatest Achievement of Chambers," <u>San Antonio Evening News</u>, March 13, 1933; <u>History of Randolph Field</u>; Rossi L. Selvaggi, "A History of Randolph Air Force Base" (University of Texas: Masters Thesis, 1958); Bruce Elton Burgoyne, "The Acquisition of Randolph Field" (Trinity University: Masters Thesis, 1957).

[47] <u>Austin American</u>, April 5, 1928, p. 1; <u>History of Randolph Field</u>, Vol. I, pp. 20-22.

[48] John J. Hibbits, <u>Take 'er Up Alone, Mister</u>, as told to F. E. Rechnitzer (New York: Whittlesey House, 1943), p. 9. The banner of the Randolph AFB home page, for example, http://www.randolph.af.mil, included the phrase "Showplace of the Air Force" throughout the time this study was written.

[49] Fred R. Freyer, <u>The Story of the 1 July 1931 Cadet Class at Brooks, March, & Randolph Fields</u> (Manhattan KS: Military Affairs/Aerospace Historian Publishing, 1975)

contains a number of cadet reminiscences, class statistics, and excerpts from the base newspapers; "Merry Christmas! From Randolph Field," The [Randolph Field] Tee, December 23, 1931; Fred R. Freyer, "My Flying Cadet Days" [1978] is a collection of letters his mother had saved from Freyer's training days, available AETC History Office; "Mess Hall Opens Cadet Xmas Gift," The Tee, January 14, 1932, pp. 3-4. The Philip quote is from Freyer, The Story of the 1 July 1931 Cadet Class, which does not have page numbers, and the Freyer quote is from his "My Flying Cadet Days," p. 24.

[50] C. R. Slawter, "World's Strictest School Trains Our War Birds: Randolph Field," Popular Science Monthly, September 1932, pp. 34-35 and 95; "Welcome Class of July 1937," The Tee, August 7, 1937, pp. 1-2; Neely C. Mashburn, "Some Interesting Psychological Factors in the Selection of Military Aviators," Aviation Medicine, December 1935, pp. 113-126; Memo, Captain W. F. Hall to Lt Colonel H. C. Davidson, [Physical Characteristics of Air Corps Training Center Graduates], February 15, 1939, AETC History Office. In oral history interviews and in the occasional writings of former aviation cadets on file in the AETC History Office, many mentioned the rigors of the physical qualification exams and how few passed.

[51] Beirne Lay, I Wanted Wings (New York: Harper & Brothers, 1937; 9th ed.). "Drill. Room inspections" from p. 6; "we all stood there" p. 47; "clanging bells" p. 52; "take it" p. 58.

[52] Ibid., "gods—every one" from p. 83, "hallowed ground" p. 144; "realism was creeping in" p. 146, and "unnecessary childishness" p. 184.

[53] Work Projects Administration, Randolph Field (New York: Devin-Adair Company, 1942), p. 43.

[54] I. H. Edwards, "To Flying Cadet Battalion," Form One, p. 2. See William H. McNeill, Keeping Together in Time: Dance and Drill in Human History (Cambridge MA: Harvard University, 1995) for a review of the history of military drill.

[55] "Wings for War," Life, January 30, 1939, pp. 45-52.

[56] "At Kelly Field" and "News of the Month at Kelly Field," The Flying Cadet, November 11, 1931, pp. 2 and 5; "long been known" and "that they were still" both from p. 5.

[57] J. B. Burwell, "Student Training at the Air Corps Training Center," Air Corps News Letter, March 1, 1937, from manuscript in the HQ AETC History Office; "The Selection and Training of Flying Cadets," Air Corps News Letter, May 15, 1937, "fundamental training and experience" at p. 9 and "we have helped" at p. 10; History of Military Aviation in San Antonio, pp. 84-86 for the types of trainer aircraft used.

[58] "The New Air Corps Act," Air Corps News Letter, July 15, 1936, pp. 10 and 19; Office of Statistical Control, "Army Air Forces Statistical Digest World War II" (Washington DC: USAAF, December 1945), p. 15.

[59] H. H. Arnold, Global Mission (New York: Harper & Brothers, 1949), p. 181; Richard F. McMullen, The Role of Civilian Contractors in the Training of Military Pilots (Scott AFB IL: Historical Division, ca. 1956); Maurice G. Stack, "The Aviation Cadet Program in Retrospect," Air University Review, July-August 1965, p. 82; History, AAFTC, 1 January 1939-7 July 1943, Vol. I, pp. 61-64, 67, 72-73; Willard Wiener, Two Hundred Thousand Flyers: The Story of the Civilian-AAF Pilot Training Program (Washington DC: Infantry Journal, January 1945) documents the history of the contract schools.

[60] History, AAFTC, 1 January 1939-7 July 1943, Vol. I, pp. 88-90; History of Military Aviation in San Antonio, pp. 90-101.

CHAPTER III

[61] The most comprehensive Air Force study of World War II is Wesley Frank Craven and James Lea Cate, eds., The Army Air Forces in World War II, 7 vols. (Washington DC: Office of Air Force History, 1983, new imprint); Part III of Volume 6 deals with recruiting and training issues.

[62] Volume I of the AAF Flying Training Command History covering the period 1 January 1939 to 7 July 1943 describes in detail the early increases in the flying training program. See also United States Statutes at Large, Vol. 55, Part 1 (Washington DC: GPO, 1942),

pp. 239-240; <u>Congressional Record</u>, Vol. 86, Part II (Washington DC: GPO, 1941), pp. 12830-12832; AAF Historical Office, <u>Legislation Relating to the AAF Personnel Program 1939-1945, Army Air Forces Historical Studies: No. 16</u> (Washington DC: Historical Office, May 1946; rev. ed.), pp. 36-39; Jensen, <u>The Aviation Cadet Ground Duty Program</u>, pp. 25-27, 54.

[63] History, AAFTC, 1 January 1939-7 July 1943, Vol. I, pp. 320-322, 339-341; Talking Paper, AETC/HO, "The Oldest MAJCOM in Today's Air Force," 25 October 94.

[64] Jensen, <u>Aviation Cadet Ground Duty Program</u>, pp. 34-35; Memo, J. W. Durant, Asst. Chief, Military Personnel Division, to E. C. Lynch, G-1 Section, "Aviation Cadet Specialized Training," 28 July 1941.

[65] "Aviation Cadet Training for the Army Air Forces," United States Army Recruiting Publicity Bureau, April 10, 1943, AFHRA call no. 520.056-423, IRIS 1076247. The October 15, 1943, "Aviation Cadet Training for the Army Air Forces" recruiting brochure is available at http://www.armyairforces.com/AAF/cadet-00.asp, accessed 6 October 2003.

[66] The AAF Flying Training Command and AAF Training Command histories from World War II detail the training programs, but do not provide complete statistical information. Jensen, <u>Aviation Cadet Ground Duty Program</u>, p. 32 notes the Aviation Cadet Section redesignation; a number of Aviation Cadet Branch weekly activity reports are available from the AFHRA under call no. 121.32, IRIS no. 111012. See <u>Army Air Forces Statistical Digest 1946</u> (Washington DC: Statistical Control Division, June 1947), p. 92 for the number of aviation cadets in training.

[67] <u>Individual Training of Navigators in the AAF, Army Air Forces Historical Studies: No. 27</u> (Washington DC: Historical Division, January 1945) has a recap of the stanine score requirements at p. 34; History, Central Flying Training Command, Medical History World War II 1940-1944, Vol. III, p. 103.

[68] See <u>Aviation Cadet Manual</u> (Governors Island: Recruiting Publicity Bureau, 1942), pp. 17 and 18 for height and weight limits; this document is included in AFHRA call no. 141.24-15, IRIS no. 114256. <u>Initial Selection of Candidates for Pilot, Bombardier, and Navigator Training</u> detailed the change in physical qualifications, see pp. 52-53. Base

newspaper articles publicized the changes in qualifications; see, for example, "Program Speeded for Air Corps," <u>Southeast Air Corps Training Center News</u>, January 3, 1942, p. 1.

[69] The number of applicants processed from W. Eugene Hollon, <u>History of Preflight Training in the AAF 1941-1953, Air Historical Studies No.90</u> (Maxwell AFB: Historical Division, June 1953), pp. 33-34. HQ AAF letter 35-66, "Suspension of Procurement of Aviation Cadets from the AAF," 8 March 1944 included in Aviation Cadet Branch weekly activity reports, AFHRA call no. 121.32, IRIS no. 111012; see also Memo, W. H. Redit, Aviation Cadet Branch, to J. W. Durant, "Historical Report, Aviation Cadet Branch, Week Ending 1 April 1944," for the exceptions for returning aircrew and African-Americans, also in AFHRA call no. 121.32, IRIS no. 111012. Charles A. Watry, a World War II-era aviation cadet explained the tests in his book <u>Washout! The Aviation Cadet Story</u> (Carlsbad: California Aero Press, 1983), pp. 48-58. See <u>Initial Selection of Candidates for Pilot, Bombardier, and Navigator Training</u>, pp. 25-26 for an explanation of the difference between the ACQE and the education test.

[70] The Psychological Division in the Office of the Air Surgeon prepared a comprehensive review of ACQE AC-10-A and changes made in subsequent versions of the test. See "A Report on the Purpose, Development and Validation of Test AC-10-A," October 1942, included in AFHRA call no. 141.24-15, IRIS no. 114256. This file also contained several versions of the test. See also <u>Initial Selection of Candidates for Pilot, Bombardier, and Navigator Training</u>, p. 45.

[71] <u>Initial Selection of Candidates for Pilot, Bombardier, and Navigator Training</u>, p. 46.

[72] Hollon, <u>History of Preflight Training</u>, p. 29 for the creation of the aircrew classification centers. See Eugene Fletcher, <u>Mister: The Training of an Aviation Cadet in World War II</u> (Seattle: University of Washington Press, 1992), pp. 20-22.

[73] Memo, FTC No. 50-25-3, "Assignment and Progress of Students," March 2, 1943, included in AFHRA call no. 220.7131-1, IRIS no. 145809, outlined the purpose and administration of the College Training Program. See also Hollon, <u>History of Preflight Training</u>, pp. 32-50, 78-79.

[74] Memo, W. H. Redit, Aviation Cadet Branch, to J. W. Durant, "Historical Report, Aviation Cadet Branch, Week Ending 22 July 1944," 25 July 1944, included in AFHRA call no. 121.32, IRIS no. 111012; J. H. MacWilliam and Bruce D. Callander, "The Third Lieutenants," Air Force Magazine, March 1990, pp. 100-102; Watry, Washout! The Aviation Cadet Story, p. 134.

[75] Ownership of Moffett Field transferred to the U.S. Navy early in 1941 and Santa Ana Field took its place as the RTC for the West Coast Air Corps Training Center according to Hollon, History of Preflight Training, p. 24.

[76] See Hollon, History of Preflight Training and Anne Hussey, Air Force Flight Screening: Evolutionary Changes 1917-2003 (Randolph AFB: AETC History Office, 2004) for histories of the preflight program. Vol. II of History, AAFTC, 1 January 1939 to V-J Day documents the changing length of the flying training program at pp. 510-512; see also p. 447 for the change in nomenclature to "Preflight School" and the creation of the separate preflight programs at pp. 447-448.

[77] Army Air Forces Statistical Digest, p. 64 for numbers of graduates; Policies and Procedures Governing Elimination from AAF Schools 1939-1945 (Maxwell AFB: USAF Historical Division, 1952), p. 4. In addition to the Training Command histories, see Rebecca Hancock Cameron, Training to Fly: Military Flight Training 1907-1945 (Washington DC: Air Force History and Museums Program, 1999) for a description of pilot training programs. The various schools are documented in Vol. 2 of the AAFTC 1 January 1939 to V-J Day history.

[78] See WPA, Randolph Field, pp. 65-85; "the damndest noise" from p. 65. Several issues of the Form One are available in the AETC History Office; see books such as Eugene Fletcher, Mister: The Training of an Aviation Cadet in World War II (Seattle: University of Washington Press, 1992) and Watry, Washout! The Aviation Cadet Story.

[79] The undated newspaper article, ca. 18 March 1943, is from a scrapbook donated to the AETC History Office by George Frame, whose father was the commandant of students at the CIS in 1943.

[80] History of Randolph Field, Vol. I, pp. 193-194; "Flying Lessons Now in Books; TC Streamlines Instruction," Maxwell Field Training News, May 6, 1944, p. 3.

[81] The AETC History Office home page, http://www.aetc.randolph.af.mil/ho, has photographs, a brief history, and operating characteristics for its trainer aircraft. See also publications such as Gordon Swanborough and Peter M. Bowers, United States Military Aircraft since 1909 (Washington DC: Smithsonian Institution Press, 1989). The TMPFF/ UMPFF anecdote is from Tom Killebrew, The Royal Air Force in Texas: Training British Pilots in Terrell During World War II (Denton: University of North Texas Press, 2003), p. 64.

[82] Bruce D. Callander, "The Aviation Cadets," Air Force Magazine, November 1990, pp. 98-101, also available from the Air Force Association web page, http://www.afa.org/magazine/1990/1190cadets.asp accessed 28 October 2003.

[83] See "Compass Headings," Flying Cadet Regiment, Randolph Field, June 20, 1941, AETC History Office and "Take-Off: Official Handbook of the Corps of Aviation Cadets," Maxwell Field, September 15, 1943, AFHRA call no. 222.716-4, IRIS no. 1076864; FTC Memorandum, No. 50-0-2, "Training: Hazing," May 15, 1943 and FTC Memorandum 50-27-6, "Training: Military Training, Delinquencies for Aviation Cadets," July 5, 1943, both included in AFHRA call no. 221.186, IRIS no. 146105.

[84] Errol D. Severe, for example, in his mostly autobiographical, The Last of a Breed (Eureka Springs AR: Lighthouse Productions, 1997), details hazing that occurred in the last years of the aviation cadet program.

[85] Memo, B. K. Yount, to Commanding General, "Hazing," March 25, 1942, included as part of AFHRA call no. 220.1681-2, IRIS no. 145484; FTC Memorandum 50-27-6, "Training: Military Training, Delinquencies for Aviation Cadets," July 5, 1943, AFHRA call no. 221.186, IRIS no. 146105. A Randolph Field cadet could be discharged if he accumulated 60 demerits while an underclassman or 40 as an upperclassman; see WPA, Randolph Field, p. 62.

[86] Quoted in WPA, Randolph Field, p. 63.

[87] Ibid.

[88] History, AAFFTC, 1 January 1939-7 July 1943, Vol. I, p. 127; History, AAFTC, 1 January 1939 to V-J Day, Vol. III, p. 935; Paper, War Department, "The Air Crew Training Courses as Bombardier or Navigator for Aviation Cadets," ca. January 1942, included in AFHRA call no. 220.1681-2, IRIS no. 145484.

[89] Part Nine of History, AAFTC, 1 January 1939 to V-J Day, Vol. III, discusses navigation training. See also Individual Training of Navigators in the AAF and Initial Selection of Candidates for Pilot, Bombardier, and Navigator Training.

[90] See Individual Training of Navigators in the AAF, pp. 119-126 for the dual training program and Part Ten of the AAFTC, 1 January 1939 to V-J Day history for coverage of the bombardier training programs; p. 936 explains the difficulty of screening for bombardiers.

[91] Memo, Wm. L. Tydings, West Coast Air Corps Training Center, to Commanding General, Air Forces Flying Training Command, "Training Morale," March 28, 1942, and AAFTC indorsement, April 17, 1942, both in AFHRA call no. 220.1681-2, IRIS no. 145484; "Bombardiers and Navigators Indispensable in Air Corps," Southeast Air Corps Training Center News, October 4, 1941, p. 2. See History, AAFTC, 1 January 1939 to V-J Day, Vol. III, pp. 937-938 for the efforts to publicize the role of the bombardier.

[92] History, AAFTC, 1 January 1939-V-J Day, Vol. V, pp. 1445-1454; History, ATC, Vol. I, 1 July-31 December 1946, pp. 167-170.

[93] History, AAF Technical Training Command, 1 January 1939-7 July 1943, Vol. III, pp. 966-983; History, AAFTC, 1 January 1939-V-J Day, Vol. V, p. 1565.

[94] Brochure, Air Corps Technical Schools, "Program of Instruction for (Non-Pilot) Aviation Cadets and Officers Courses, School Year 1941-1942," December 1941; History, AAF Training Command, 1 January 1939-V-J Day, Vol. V, pp. 1565-1567, 1591-1598.

[95] Jensen, Aviation Cadet Ground Duty Program, is the best single source for the ground duty programs. See also Craven and Cate, The Army Air Forces in World War II, Vol. 6, pp. 445-450. See Aviation Cadet Manual pp. 17 and 18 for height and weight limits.

[96] Jensen, <u>Aviation Cadet Ground Duty Program</u>, pp. 40-42; "Aviation Cadet Training for Enlisted Men," <u>Air Forces General Information Bulletin</u>, February 1943, pp. 32-33, AFHRA call no. 142.0372, IRIS no. 115123.

[97] Jensen, <u>Aviation Cadet Ground Duty Program</u>, pp. 82-84; <u>Army Air Forces Statistical Digest, Supplement Number 1, 1945</u> (Washington DC: Office of Statistical Control, April 1946), p. 23.

[98] Dominick A. Pisano, <u>To Fill the Skies with Pilots: The Civilian Pilot Training Program, 1939-1946</u> (Urbana: University of Illinois Press, 1993) is a history of the CPT program; see p. 75 for the schools that offered programs for African-American students.

[99] Books about the Tuskegee Airmen include <u>The Tuskegee Airmen: The Story of the Negro in the U.S. Air Force</u>, written by Charles E. Francis, one of the African Americans trained at Tuskegee. The first edition of this book was published by Bruce Humphries Inc., Boston, in 1955. See also Alan M. Osur, <u>Blacks in the Army Air Forces during World War II</u> (Washington DC: Office of Air Force History, 1977); Robert J. Jakeman, <u>Divided Skies: Establishing Segregated Flight Training at Tuskegee, Alabama, 1934-1942</u> (Tuscaloosa AL: University of Alabama, 1992); Morris J. MacGregor, Jr., <u>Integration of the Armed Forces 1940-1965</u> (Washington DC: Center of Military History, 1981); and Alan L. Gropman, <u>The Air Force Integrates 1945-1964</u> (Washington DC, 1978). There are a number of references available on the internet. See, for example, the Tuskegee Airmen Incorporated homepage at http://www.tuskegeeairmen.org and the National Park Service "Tuskegee Legends" page at http://www.cr.nps.gov/museum/exhibits/tuskegee/index.htm, both accessed January 2004. A discussion about whether African-Americans would qualify for the aviation cadet programs can be found in <u>Congressional Record</u>, Vol. 86, Part II, pp. 12831-12832. An abbreviated history of the 477th Composite Group can be found in <u>Air Force Combat Units of World War II</u>, Maurer Maurer ed. (Washington DC: Office of Air Force History, 1983), pp. 349-350.

[100] <u>Army Air Forces Statistical Digest 1946</u>, p. 75; Osur, <u>Blacks in the Army Air Forces during World War II</u>, pp. 39-42; MacGregor, <u>Integration of the Armed Forces</u>, p. 271; Gropman, <u>The Air Force Integrates 1945-1964</u>, pp. 7, 14-15. See Vol. II of History, AAFTC, 1 January 1939 to V-J Day, p. 387 for information about the stanines used for aircrew classification.

[101] Osur, <u>Blacks in the Army Air Forces during World War II</u>, pp. 31, 53; <u>Army Air Forces Statistical Digest 1946</u>, p. 75.

[102] The Army Air Forces History Office reviewed the contributions of women pilot to the war effort in its study <u>Women Pilots with the AAF, 1941-1944</u>, completed in 1946; see also Memo to Commanding General, Army Air Forces, "Women Airforces Service Pilots," 1 August 1944. A number of former WASP have written about their experiences, and Texas Woman's University has an extensive WASP-related archive and an is actively conducting oral history interviews with surviving WASP. Dorothy Schaffter examined the varied roles of women in the armed forces during the war and the potential social implications for the American Council on Education; see <u>What Comes of Training Women for War</u> (Washington DC: American Council on Education, 1948). Sanders, <u>Legislation Relating to the AAF Personnel Program 1939-1945</u>, discussed the various proposals to make the women pilots a part of the military; see pp. 24-29.

[103] History, AAFTC, 1 January 1939-V-J Day, Vol. I, pp. 42-43 for the list of bases as of 31 August 1945 and Vol. II, pp. 449-450 for the preflight consolidation and moves; History, AAFTC, 1 May 1945-V-J Day, pp. 30-39, 70-86; History, AAFTC, 1 September 1945-31 December 1945, pp. 23-24; "Pre-Flight School at Maxwell Closes," <u>Training News</u>, December 2, 1944, p. 1; <u>Army Air Forces Statistical Digest 1946</u>, p. 75.

CHAPTER IV

[104] The three-month long OTS program, which prepared college graduates to be officers, eliminated the need for OCS, with its six-month long preparation course for non-college grads. OTS not only supported the goal of creating an all college-graduate officer corps, it also reduced training costs and increased the flexibility of production goals by reducing the time needed to ready new officers for service. See Thomas A. Manning, et. al, <u>History of Air Training Command 1943-1993</u> (Randolph AFB: AETC History Office, 1993), pp. 133, 150-151 for short histories of OTS and OCS.

[105] History, AAFTC, 1 September-31 December 1945, pp. 23-24, 31-34, 41.

[106] Ibid., pp. 70-83.

[107] Vance O. Mitchell, <u>Air Force Officers Personnel Policy Development 1944-1974</u> (Washington DC: Air Force History and Museums Program, 1996), pp. 55-57, 107-116. Officer Training School, implemented in 1959, also contributed to the pool of non-rated officers; see Mitchell, p. 166-167.

[108] Email, LewJtn@SMTP to Av1cadet@SMTP, "Latest History," June 30, 1996; Special Order Number 249, Williams Field, 18 October 1946, courtesy Lew Johnston, Aviation Cadet Alumni Association, San Francisco.

[109] History, AAFTC, 1 January-30 June 1946, pp. 9, 41-47.

[110] The Army redesignated AAFTC as Air Training Command on 1 July 1946. Earlier in the year, command headquarters moved from Fort Worth, Texas, to Barksdale Field, Louisiana. In 1949, command headquarters moved to Scott AFB, where it would reside until 1957 when it moved to its present location at Randolph AFB.

[111] Ibid., pp. 66-69; History, ATC, 1 July-31 December 1946, pp. 159-160.

[112] History, ATC, 1 January-31 December 1947, pp. 64-67, 234; History, Flying Division, 1 October-31 December 1946, Vol. I, pp. 98-101; History, Flying Division, 1 January-31 March 1947, pp. 105, 110-111, 349-350.

[113] History, ATC, 1 January 1948-30 June 1949, pp. 44-53, 67, 78.

[114] Ibid., pp. 55-61, 83-84; History, ATC, 1 July-31 December 1949, pp. 79-89.

[115] In January 1948, just months after attaining its independence, the Air Force changed the designation of its major installations to "Air Force Base" instead of "Army Air Field."

[116] History, ATC, 1 January 1948-30 June 1949, pp. 88-95; History, ATC, 1 July-31 December 1949, pp. 147-160, 172-174.

[117] History, ATC, 1 July-31 December 1949, pp. 155-160, 181-182; History, 1 January-30 June 1950, Vol. I, pp. 172-173.

[118] History, ATC, 1 July 1950-30 June 1951, Vol. I, pp. 130-132, 137-141, 157-158; Daniel L. Haulman, "Air Force Bases, 1947-1960," Chapter 2 of History's Footprints: USAF Continental Bases, Frederick J. Shaw, ed. (Maxwell AFB: AFHRA, 2004, draft ms.), pp. 112-115.

[119] History, ATC, 1 July 1950-30 June 1951, Vol. I, pp. 171-174.

[120] Ibid., pp. 176-180.

[121] Memo, [ATC] Personnel Planning to [ATC] Personnel, "Aviation Cadet Selection Criteria," 31 July 1951, SD II-1 in History, Air Training Command, July 1950-June 1951, Vol. II; Paper, ATC, "Aviation Cadet Procurement."

[122] History, FTAF, 1 January-30 June 1952, Vol. I, pp. 102-106; Paper, ATC, "Aviation Cadet Procurement," Personnel Planning Project Number 20-9-52, October 1952, available at the Air University Library; "Aviation Cadets," Air Training, June 1953, p. 6. See also Mitchell, Air Force Officers Personnel Policy Development, p. 105.

[123] History, ATC, 1 July-31 December 1953, pp. 79, 96-101.

[124] History, ATC, 1 July-31 December 1952, pp. 35-43, 50-52; History, ATC, 1 July-31 December 1953, p. 99; Hollon, History of Preflight Training, pp. 182-190. The preflight training syllabus is Supporting Document II-6 in the 1 January-30 June 1953, FTAF history.

[125] History, ATC, 1 July-31 December 1952, pp. 35-43, 50-52; History, ATC, 1 July-31 December 1953, p. 99; Hollon, History of Preflight Training, pp. 182-190.

[126] History, ATC, 1 July-31 December 1952, p. 36; History, FTAF, 1 January-30 June 1954, Vol. I, p. 89; History, FTAF (FOUO), 1 January-30 June 1956, Vol. I, p. 79, information used not FOUO.

[127] "Ramp Out, Class 51-H," Reese AFB.

[128] History, ATC, 1 July-31 December 1952, pp. 52-53.

[129] The three commands were Flying Training Air Force, Technical Training Air Force (TTAF), and Crew Training Air Force (CTAF). FTAF was activated on 1 May 1951, TTAF on 16 July 1951, and CTAF on 16 March 1952. In 1957, the command inactivated CTAF, transferring its mission to FTAF; FTAF and TTAF were inactivated in 1958, with their missions reverting to HQ ATC.

[130] History, FTAF, 1 January-30 June 1952, Vol. I, pp. 280-369, passim.; History, ATC, 1 July-31 December 1953, p. 116; Hollon, History of Preflight Training, p. 188.

[131] History, FTAF, 1 January-30 June 1952, Vol. I, p. 392.

[132] History, 3500th USAF Recruiting Squadron (Aviation Cadet), 1 April-30 June 1952, pp. 1-7, AFHRA Call No. K-SQ-RCTC-3500-HI, IRIS No. 431507; History, FTAF, 1 January-30 June 1953, pp. 107-110; History (FOUO), FTAF, 1 July-31 December 1953, Vol. I, p. 93, information used not FOUO.

[133] History, ATC, 1 July-31 December 1952, Vol. I, pp. 58-64; History, FTAF (FOUO), 1 July-31 December 1952, Vol. I, pp. 73-74, information used not FOUO.

[134] Johnson Poor, "Where Tigers Learn to Growl," Air Training, October 1953, p. 10; Pamphlet, "The Training of an Air Force Officer," September 20, 1954. The tiger roars, apparently, were a short-lived phenomenon. In a discussion with the author in February 2004, Clem Bellion, who completed preflight in August 1955, did not remember having to roar or growl during his training. The idea that "every man was a tiger," however, was an important concept, and one of the photographs of Bellion in his scrapbook is labeled "First Tiger Picture."

[135] History, FTAF, 1 January-30 June 1953, Vol. I, pp. 29-37.

[136] History, FTAF (FOUO), 1 July-31 December 1953, Vol. I, pp. 63-64, information used not FOUO; History (FOUO), FTAF, 1 January-30 June 1955, Vol. I, p. 158, information used not FOUO.

[137] History, FTAF, 1 January-30 June 1955, Vol. I, p. 158-161; History (FOUO), FTAF, 1 January-30 June 1955, Vol. I, pp. 136-138, information used not FOUO.

[138] History, ATC, 1 July-31 December 1954, Vol. I, pp. 253-255.

[139] Briefing, ATC Management Analysis Directorate, "A Study of Aviation Cadet Attrition," June 1959, SD 63 in History, ATC, 1 January-30 June 1959, Vol. III; Message, HQ USAF to ALMAJCOM, "Aviation Cadet Testing," 21 October 1959, SD 158 in History, ATC, 1 July-31 December 1959, Vol. V; History (S), ATC, 1 July 1961-31 December 1961, Vol. I, p. 95, information used not classified.

[140] History, FTAF, 1 January-30 June 1954, Vol. I, pp. 90-91; Point Paper, Ms. Parrish, AETC/HO, "Recruiting Service Emblem," 24 Oct 95.

[141] History, FTAF, 1 January-30 June 1955, Vol. I, pp. 99-100.

[142] Mike Worden, Rise of the Fighter Generals: The Problem of Air Force Leadership 1945-1982 (Maxwell AFB: Air University Press, March 1998), p. 88.

[143] History (FOUO), FTAF, 1 July-30 December 1952, Vol. I, p.78, information used not FOUO; History (FOUO), ATC, 1 January-30 June 1961, Vol. I, p. 91, information used not FOUO; Robert C. Bueker, "The Last Cadet," Airman, May 1965, p. 31; George F. Lemmer, USAF Manpower Trends 1960-1963 (Washington DC: USAF Historical Division Liaison Office, March 1965), p. 34. Mitchell estimated that perhaps as few as 12 percent of rated officers received their commissions via the aviation cadet program by 1959; see Mitchell, Air Force Officers Personnel Policy Development, p. 161.

[144] Memo, ATFTM-P/Mr. Lilly to Commander, "Success of 1959 Academy Graduates," 21 April 1960, SD 198, History, ATC, 1 January-30 June 1960, Vol. V; Report, Training Evaluation Division, "An Evaluation of the 1960 Academy Graduates in Pilot Training," October 1961, SD III-13, History, ATC, 1 July-31 December 1961, Vol. XIII.

[145] History, ATC (FOUO), 1 January-30 June 1961, Vol. I, pp. 17-20, 99-100, material used not FOUO. Severe, Last of a Breed, provides the reader a detailed account of life as an aviation cadet from preflight to commissioning in 1960-1961.

[146] Hussey, Air Force Flight Screening, pp. 12, 16; History of Air Training Command 1943-1993, p. 142.

[147] History, ATC (FOUO), 1 January-30 June 1961, Vol. I, pp. 115-119, information used not FOUO.

[148] Paper, Thomas R. Carretta, "Sex Differences on U.S. Air Force Pilot Selection Tests," from the Air Force Research Laboratory home page, http://www.hec.afrl.af mil/publications/ISAP_97R.doc.pdf accessed 15 December 2003.

[149] Memo, ATOFT to ATOFT-P, "Elimination Rate, Aviation Cadet Preflight, Pilot and Navigator," 20 Oct 1958, SD 51 in History, ATC, 1 July-31 December 1958, Vol. III; Message, COMATC to CofSUSAFWASHDC, [Shortage of Pilot Aviation Cadet Applicants], 262219Z Nov 58, SD 56 in History, ATC, 1 July-31 December 1958, Vol. III; Message, HEDUSAF to COMATC, [Shortage of Pilot Aviation Cadet Applicants], 152217Z Dec 58, SD 57 in History, ATC, 1 July-31 December 1958, Vol. III; Memo, USAF/AFPTR-F to ATC/CC, "Reduction in the USAF Pilot Training Rate," 15 Jan 59, SD 32 in History, ATC, 1 January-30 June 1959, Vol. III.

[150] Message, HQ USAF to ATC, "Preflight Schools," 191947Z Jan 60, SD 31 in History, ATC, 1 January-30 June 1960, Vol. III; History (S NF FOUO), ATC, 1 July-31 December 1959, pp. 58-59, information used not classified or FOUO; Memo, ATC/ATPOP-PA to All Air Training Command Activities less USAF Recruiting Service, "Aviation Cadet Procurement Program," 28 Jul 59, SD 146 in History, ATC, 1 July-31 December 195, Vol. IV; Message, HQ USAF to ATC, [Aviation Cadet Recruiting], 231914Z Dec 59, SD 160 in History, ATC, 1 July-31 December 1959, Vol. IV; History (S NF RD), ATC, 1 January-30 June 1960, pp. 57, 61, information used not classified or RD; Message, HQ USAF to ATC, "Preflight Schools," 191947Z Jan 60, SD 31 in History, ATC, 1 January-30 June 1959, Vol. III; Memo, USAF/AFPTR to ATC, et al., "Flying Training Program Guidance for PFT 62-1," 10 Feb 60, SD 176 in History, ATC, 1 January-30 June 1960, Vol. V.

[151] Severe, Last of a Breed, pp. 3-6.

[152] History, ATC, 1 January-30 June 1964, pp. 146, 148; Message, ATC to USAF Rectg Svc, [Aviation Cadet Recruiting], 18 Oct 63, SD II-4 in History, ATC, 1 July-31 December 1963; Memo, ATC/CV to USAF Recrtg Svc, "Status of OTS Procurement," 26 Jul 63, SD II-5, in History, ATC, 1 July-31 December 1963.

[153] Bueker, "The Last Cadet."

CONCLUSION

[154] Report, <u>Rand Symposium on Pilot Training and the Pilot Career: Final Report</u> (Santa Monica CA: Rand, December 1970), p. 3; report available from the Air University Library.

[155] Ibid., pp. 12-19; History (S), ATC, FY 1971, Vol. I, pp. 166-168, information used not classified.

[156] Maurice G. Stack, "The Aviation Cadet Program in Retrospect," <u>Air University Review</u>, July-August 1965, pp. 78 and 90.

[157] Mitchell, <u>Air Force Officers Personnel Policy Development</u>, pp. 301-302.